Diction
Quantities a

DICTIONARY OF QUANTITIES AND UNITS

J. V. Drazil

LEONARD HILL
LONDON 1971

An Intertext Publisher

First published 1971 by
Leonard Hill Books
A division of
International Textbook Company Ltd
158 Buckingham Palace Road, London SW1

ISBN 0 249 44092 X

Printed in Great Britain at
The Pitman Press, Bath

Contents

Foreword

This book is entitled Dictionary for the sake of simplicity. The term should simply imply that the various entries are arranged in alphabetical order. The book contains basically lists of (1) names of units, (2) symbols (and abbreviations) of units, (3) names of quantities, and (4) symbols of quantities, which are arranged and cross referenced in such a way that if the reader knows one of the above mentioned four information elements he can find the remaining three. With the help of the Indices at the end of this book he can do so even if he knows one of these names only in French or German, so that this Dictionary can be successfully used by any French or German speaking person having only an elementary knowledge of English. More about the information contained in this book is explained in the Introduction and is apparent from the Contents.

With some exceptions, units which are of historical interest only, and also units which are not or cannot be converted into SI units (such as various units used in hydrometry, viscosimetry, for measuring hardness, photographic emulsion speed etc.) are not included.

Recommended multiples and sub-multiples of various units are apparent from the examples under individual headings of the relevant prefixes in Part I of this Dictionary.

Great care has been taken to ensure that the names of quantities in English, French and German correspond to those recommended by international and national standards. Some alternative or obsolescent terms are included to assist readers who may meet them in literature or who are not acquainted with the latest trends in this field. Throughout the Dictionary the English spelling (which is also the spelling used in ISO and IEC publications) is used, and the adjectives in all German names of quantities are written with lower case letters.

The publishers and author will be grateful for any suggestions for improvement of this Dictionary.

Introduction

(a) SYSTEMS AND UNITS

(1) *SI system and SI units*

The SI system (International System of Units; Système International d'Unités; Internationales Einheitensystem) is based on six base-units (unités de base; Basiseinheiten), formerly called fundamental units, which are the units of six base-quantities (grandeurs de base; Basisgrössen) listed below:

BASE-QUANTITY	BASE-UNIT	SYMBOL
length	metre	m
mass	kilogramme	kg
time	second	s
electric current	ampere	A
thermodynamic temperature	kelvin	K
luminous intensity	candela	cd

The SI system further comprises two so-called supplementary units (radian and steradian) and a number of derived units. The supplementary and derived units having special names and symbols and approved by CGPM are:

QUANTITY	UNIT	SYMBOL
plane angle	radian	rad
solid angle	steradian	sr
frequency	hertz	Hz
force	newton	N
work, energy, quantity of heat	joule	J
power	watt	W
quantity of electricity	coulomb	C
electric potential, potential difference, tension, electromotive force	volt	V
electric capacitance	farad	F
electric resistance	ohm	Ω

QUANTITY	UNIT	SYMBOL
flux of magnetic induction, magnetic flux	weber	Wb
magnetic flux density, magnetic induction	tesla	T
inductance	henry	H
luminous flux	lumen	lm
illumination	lux	lx

All the above units are indicated in this dictionary as SI units, and their multiples or sub-multiples as mSI units.

(2) *Additional units*

Other units which have special names and may be used in addition to SI units are the following:

QUANTITY	UNIT	SYMBOL
plane angle	degree	....°
	minute	....′
	second	....″
	grade	....^{g}
area	are	a
	hectare	ha
volume	litre	l
time	day	d
	hour	h
	minute	min
mass	tonne	t
pressure	bar	bar
dynamic viscosity	centipoise	cP
kinematic viscosity	centistokes	cSt
energy, work	kilowatt hour	kWh
	electronvolt	eV
Celsius temperature	degree Celsius	°C
conductance	siemens	S
reactive power	var	var
amount of substance	mole	mol

These additional units which may be used are indicated in this dictionary as oSI units, and those which are expected to become SI units are indicated as (SI) units.

They include multiples and sub-multiples of litre (hl, cl, ml) and bar (hbar, mbar, μbar).

Other tolerated units include angström, atomic mass unit (unified), barn, curie, decibel, neper, rad and röntgen.

(3) *Other systems and units*

Apart from SI units more important units of the following systems are included in this dictionary:

(*A*) Systems of units of mechanics

(1) Absolute or LMT systems (base-quantities: length, mass, time)
MKS system (base-units: metre, kilogramme, second)
MTS system (base-units: metre, tonne, second)
CGS system (base-units: centimetre, gramme, second)
FPS system (base-units: foot, pound, second)

(2) Gravitational or LFT systems (base-quantities: length, force, time)
m-kgf-s system (base-units: metre, kilogramme-force, second)
m-kp-s system (base-units: metre, kilopond, second)
ft-lbf-s system (base-units: foot, pound-force, second)

(*B*) Systems of units of electricity and magnetism
MKSA system (base-units: metre, kilogramme, second, ampere)
CGSm system (electromagnetic CGS system)
CGSe system (electrostatic CGS system)
CGSF system (base-units: centimetre, gramme, second, franklin)
CGSB system (base-units: centimetre, gramme, second, biot)

Units of other systems and units which do not belong to any system are also included in this dictionary.

Note:

(*a*) MKS and MKSA systems (and also MKS°K and m-sr-s-cd systems) form part of the SI system and are therefore not referred to separately.

(*b*) m-kgf-s and m-kp-s systems are identical (they only use a different name for the same unit of force), and m-kp-s system is therefore not referred to separately; m-kgf-s units can be changed to m-kp-s units by simply substituting kp for kgf.

(b) DESCRIPTION OF ENTRIES IN PARTS I AND II

The entry of a unit comprises:

(1) the *name* of the unit

(2) the *symbol* used for the unit, which is either

  (*a*) an *internationally used symbol* for the unit (if there is such a symbol), or

  (*b*) a *commonly used symbol* for the unit. The latter symbol is always in parentheses to distinguish it from the former one. If

  (*c*) *no symbol* is used for the unit, a dash in parentheses (—) follows the name of the unit

(3) the *French* and *German names* of the unit. This applies only to the more important units having their own name. The names of other units can be derived from the table *Expressions used in the names of units* on page 6 and the table *Prefixes and their symbols* . . . on page 7

(4) the *system* or *country* in which the unit is used. If a unit is used in more than one system, only the most important system is mentioned

(5) a *comment*, such as obsolete or deprecated, where appropriate

(6) the *quantity* or *quantities* for which the unit is used. Only one quantity is given for non-SI units

(7) the *relation* of the unit to other important units and/or *conversion factor*(*s*) for conversion of the unit to other unit(s) for the same quantity. In the case of non-SI units the first conversion factor is for conversion to the appropriate SI unit. All conversion factors printed in bold type are exact values
(8) a *note* (where appropriate)

The entry of a symbol or an abbreviation of a unit comprises:

(1) the *symbol* or *abbreviation*
(2) the *country* in which the abbreviation is used and the *full name* in the language of the country (where appropriate)
(3) a *comment*, such as obsolete or deprecated, where appropriate
(4) the *name* of the unit (in English) for which the symbol or abbreviation is used
(5) the *correct symbol* in cases where the symbol under (1) is an incorrect one.

The entry of a quantity or constant comprises:

(1) the *name* of the quantity or constant in *English*, *French* and *German*. The authenticity of the French and German names has been ensured by consulting corresponding ISO publications in French, French and German standards and modern books on physics (see Bibliography)
(2) the internationally recommended *symbol* for the quantity or constant. Symbols which are not internationally recognized are in parentheses
(3) the *dimension* of the quantity or constant
(4) the latest accepted *value* for the constant without the standard deviation error
(5) *SI unit* for the quantity or constant
(6) *other unit* which may be used instead of the SI unit
(7) a *note* (where appropriate).

(c) EXPRESSIONS USED IN THE NAMES OF UNITS

English	French	German
square ...	... carré(e)	Quadrat...
... squared	... carré(e)	Quadrat...; (... hoch zwei)
cubic ...	... cube	Kubik ...
... cubed	... cube	Kubik...; (... hoch drei)
... to the fourth	... bicarré(e) ... à la puissance quatre	... hoch vier
... to the fifth	... à la cinquième puissance	... hoch fünf
reciprocal ...	par ...; ... inverse	reziproke(-r, -s) ...
... per ...	... par ...	... durch ...

Note:

(*a*) The terms '... hoch zwei' and '... hoch drei' are rarely used (cf. DIN 1301 Entwurf—1969)

(*b*) The terms 'l'inverse du centimètre' and 'la seconde à la puissance moins un' are used in the French Standard FD X 02–002 (1966), but ISO publications written in French normally use 'par ...' (e.g. par seconde) and exceptionally '... inverse' (e.g. henry inverse)

(*c*) The German Committee for Units (Ausschuss für Einheiten und Formelgrössen im Deutschen Normenausschuss) recommends that only '... durch ...' be used in the names of units, and not '... je ...' which was used until recently in German Standards, or '... pro ...' used in older books.

(d) PREFIXES AND THEIR SYMBOLS FOR DECIMAL MULTIPLES AND SUB-MULTIPLES OF UNITS

Prefix			Symbol	Factor by which the unit is multiplied
English	French	German		
tera	téra	Tera	T	10^{12}
giga	giga	Giga	G	10^{9}
mega	méga	Mega	M	10^{6}
kilo	kilo	Kilo	k	10^{3}
hecto	hecto	Hekto	h	10^{2}
deca	déca	Deka	da	10
deci	déci	Dezi	d	10^{-1}
centi	centi	Zenti	c	10^{-2}
milli	milli	Milli	m	10^{-3}
micro	micro	Mikro	μ	10^{-6}
nano	nano	Nano	n	10^{-9}
pico	pico	Piko	p	10^{-12}
femto	femto	Femto	f	10^{-15}
atto	atto	Atto	a	10^{-18}

Note:

(*a*) The prefixes can be used both for SI and non-SI units

(*b*) The prefixes (symbols) are directly attached to the names of units (symbols of units)

(*c*) Multiples and sub-multiples of units formed by the prefixes (symbols) can be raised to a positive or negative power, e.g. $1 \text{ mm}^2 = (10^{-3} \text{ m})^2 = 10^{-6} \text{ m}^2$

(*d*) Prefixes hecto, deca, deci and centi should be avoided as far as possible

(*e*) No other prefixes or compound prefixes (e.g. decimilli) are used.

(e) SYMBOLS AND ABBREVIATIONS USED IN THIS DICTIONARY

(1) *Mathematical symbols*

.	decimal point (e.g. 3.141 59)
$\times$	multiplied by (e.g. 3.6×10^3)
/	divided by; per (e.g. 1/16 or $1/c^2$ or m/s)
=	(exactly) equal to (e.g. N = m kg/s^2; exact numerical values are printed in bolt type, e.g. 1 ft = $\mathbf{3.048 \times 10^{-1}}$ m)
$\approx$	approximately equal to (e.g. $1/3 \approx 3.333\,33 \times 10^{-1}$)
$\triangleq$	corresponds to (e.g. 1 sA $\triangleq 3.335\,64 \times 10^{-10}$ A)

Note:

(*a*) decimal comma (,) is used in non-English texts (e.g. 3,141 59)

(*b*) dot (.) or centred dot (·) are sometimes used in non-English texts instead of the symbol × for multiplication (e.g. $3{,}6 . 10^3$ or $3{,}6 \cdot 10^3$ or kg.m or kg·m). The mere positioning of symbols next to each other is often used for multiplication both in English and non-English texts; the latter method is used in this dictionary for multiplication and dimensional symbols (e.g. Nm or kg m or LM)

(*c*) In this dictionary all symbols behind the solidus form part of the denominator [e.g. $m^2/Vs = m^2/(V \cdot s)$]. Sometimes inverse power is used to indicate that a symbol is in the denominator [e.g. $m^{-1} = 1/m$, $m^2kg\,s^{-3}A^{-1} = m^2 \cdot kg/(s^3 \cdot A)$, $L^2MT^{-2}I^{-2} = L^2 \cdot M/(T^2 \cdot I^2)$]

(*d*) The last three of the above symbols (=, ≈, ≙) are omitted in front of a conversion factor at the beginning of a line.

(2) *General abbreviations*

abbr.	abbreviation
cf.	compare
Def	definition
depr.	deprecated
Dim	dimension
e.g.	for example
F	French

G	German
i.e.	that is
obsol.	obsolete; obsolescent
q.v.	which see
UK	United Kingdom
US	United States of America

(3) *Relationship of units to systems and countries*

CGS unit	unit of the cm-g-s system
CGSB unit	unit of the cm-g-s-Bi system
CGSe unit	unit of the electrostatic CGS system; esu unit
CGSF unit	unit of the cm-g-s-Fr system
CGSm unit	unit of the electromagnetic CGS system; emu unit
FPS unit	unit of the ft-lb-s system
ft-lbf-s unit	unit of the ft-lbf-s system
m-kgf-s unit	unit of the m-kgf-s system
m-kp-s unit	unit of the m-kp-s system
MTS unit	unit of the m-t-s system
mSI unit	unit which is a multiple or sub-multiple of an SI unit
oSI unit	other unit which can be used in addition to SI units
SI unit	unit of the SI system (International System of Units)
(SI) unit	unit which will probably become an SI unit in the future
UK unit	unit used in the United Kingdom
US unit	unit used in the United States of America

Note:

(*a*) When a unit is a unit of the SI system and also of some other system, the other system is mentioned only exceptionally

(*b*) Units used only in France or only in Germany include in their entries a note without any abbreviations.

(4) *Other useful abbreviations*

AEF	Ausschuss für Einheiten und Formelgrössen im Deutschen Normenausschuss
AFNOR	Association Française de Normalisation
ASA	American Standards Association
BIPM	Bureau International des Poids et Mesures
BS	British Standard
BSI	British Standards Institution
CGPM	Conférence Générale des Poids et Mesures
CIPM	Comité International des Poids et Mesures
DIN	Deutsche Industrienorm
DNA	Deutscher Normenausschuss
FR	Federal Register (US)
IAU	International Astronomical Union
ICPS	International Conference on Properties of Steam
IEC	International Electrotechnical Commission
ISO	International Organization for Standardization
IUPAC	International Union of Pure and Applied Chemistry
IUPAP	International Union of Pure and Applied Physics
WMA	Weight and Measures Act (UK)

Acknowledgements

I wish to express my thanks to my friend and colleague Mr. T. C. Wood for his co-operation and valuable advice during work on the manuscript. It was he who suggested that I should write this Dictionary and put me into contact with the Publishers, thus being partly responsible for its existence.

Much information in this book has been based on data obtained from British Standards and ISO Recommendations (see Bibliography) by permission of British Standards Institution, 2, Park Street, London W1A 2BS from whom copies of the complete publications may be obtained.

The librarians of the Buckinghamshire County Library in Amersham, of the British Standards Institution Library, of the Library of the Institution of Electrical Engineers and of the National Reference Library of Science and Invention deserve my special gratitude for their assistance.

Miss Ann Drybrough-Smith, chief editor of the Publishers, has been most helpful in advising and guiding me on the way from the first draft of the Dictionary to its present form.

Finally I would like to thank my daughter Alice for her help with the preparation of the French and German indices.

PART I

A Dictionary of Units

Their Symbols and Abbreviations

a are; year; atto; ab

A ampere; atta

Å ångström

aA abampere

ab (a) a prefix denoting a CGSm unit (used in US)

abampere (aA)
CGSm unit of electric current
10 A

abampere centimetre squared ($aA\ cm^2$)
CGSm unit of electromagnetic moment
10^{-3} Am^2

abampere per square centimetre (aA/cm^2)
CGSm unit of current density
10^5 A/m^2

abcoulomb (aC)
CGSm unit of electric charge and electric flux
For electric charge:
10 C
For electric flux:
$7.957\,75 \times 10^{-1}$ C

abcoulomb centimetre (aC cm)
CGSm unit of electric dipole moment
10^{-1} C m

abcoulomb per cubic centimetre (aC/cm^3)
CGSm unit of volume density of charge
10^7 C/m^3

abcoulomb per square centimetre (aC/cm²)

CGSm unit of surface density of charge, electric polarization and displacement

For surface density and polarization:

10^5 C/m²

For displacement:

$7.957\,75 \times 10^3$ C/m²

abfarad (aF)

CGSm unit of capacitance

10^9 F

abhenry (aH)

CGSm unit of self inductance, mutual inductance and permeance

For inductances:

10^{-9} H

For permeance:

$1.256\,64 \times 10^{-8}$ H

abmho (1/aΩ)

= absiemens (q.v.)

abohm (aΩ)

CGSm unit of resistance

10^{-9} Ω

abohm centimetre (aΩ cm)

CGSm unit of resistivity

10^{-11} Ω m

absiemens (aS)

CGSm unit of conductance

10^9 S

absiemens per centimetre (aS/cm)

CGSm unit of conductivity

10^{11} S/m

absolute

Note: depr. adjective sometimes used to distinguish MKSA (= SI) units from the so-called international units (which are obsol. since 1947), so that e.g. the symbol A_{abs} is used for the ampere (A) to distinguish it from the international ampere (A_{int})

abstat depr. prefix denoting a CGSe unit

abvolt (aV)

CGSm unit of electric potential

10^{-8} V

abvolt per centimetre (aV/cm)

CGSm unit of electric field strength

10^{-6} V/m

aC abcoulomb

acoustic ohm (—)

unit of acoustic impedance

10^5 N s/m^5

Note: name sometimes given to dyne second per centimetre to the fifth

acre (—)

UK and US unit of area

4.046 86 × 10^3 m^2

4.046 86 × 10^{-1} ha

1.562 5 × **10^{-3}** mile^2

4.840 × **10^3** yd^2

Note: used for agrarian measurements

acre-foot (acre ft)

US unit of volume

1.233 48 × 10^3 m^3

4.356 × **10^4** ft^3

Note: used in irrigation engineering

acre foot per day (acre ft/d)
US unit of volume rate of flow
$1.427\,64 \times 10^{-2}$ m^3/s
$1.233\,48 \times 10^{3}$ m^3/d
4.356 $\times$ **10^4** ft^3/d

acre foot per hour (acre ft/h)
US unit of volume rate of flow
$1.233\,48 \times 10^{3}$ m^3/h
4.356 $\times$ **10^4** ft^3/h

acre-inch (acre-in)
US unit of volume
$1.027\,90 \times 10^{2}$ m^3
3.630 $\times$ **10^3** ft^3

acre per pound (acre/lb)
ft-lbm-s unit of specific surface
$8.921\,79 \times 10^{-1}$ ha/kg

admiralty mile obsol. name for nautical mile (UK)

AE G symb. for *astronomische Einheit* = astronomical unit

aF abfarad

A.G. G. abbr. for *Atomgewicht* = atomic weight

ah; a-h ampere hour (Ah)

aH abhenry

Ah ampere hour

a.l. F abbr. for *année de lumière* = light year

amp, Amp ampere (A)

ampere A

ampère

Ampere

SI base-unit of electric current and SI unit of magnetic potential difference and magnetomotive force

Def: the ampere is the unit of electric current defined under that name by the CGPM

$2.997\,92 \times 10^{9}$ sA or Fr/s

10^{-1} aA or Bi

ampere × circular mil (A × circular mil)

unit of electromagnetic moment

$5.067\,07 \times 10^{-10}$ Am^2

ampere hour Ah unit of electric charge

3.6×10^{3} C or As

ampere metre squared Am^2

SI unit of electromagnetic moment (magnetic moment)

$2.997\,92 \times 10^{13}$ sA cm^2

10^{3} aA cm^2 or Bi cm^2

Note: also unit of Bohr magneton and nuclear magneton

ampere minute A min

unit of electric charge

$1.666\,67 \times 10^{-2}$ Ah

ampere per kilogramme A/kg

= coulomb per kilogramme second (q.v.)

ampere per metre A/m

SI unit of magnetic field strength and magnetization = N/Wb

For magnetic field strength:

$1.256\,64 \times 10^{-2}$ Oe or Gb/cm

$3.767\,30 \times 10^{8}$ CGSe unit

For magnetization:

$3.335\,64 \times 10^{-7}$ CGSe unit

$1.000\,00 \times 10^{3}$ CGSm unit

ampere per square inch A/in^2
unit of current density
1.550 00 × 10^3 A/m^2

ampere per square metre A/m^2
SI unit of current density
2.997 92 × 10^5 sA/cm^2
10^{-5} aA/cm^2

ampere per volt A/V
= siemens (q.v.)

ampere per weber A/Wb
= reciprocal henry (q.v.)

ampere second As
= coulomb (q.v.)

ampere square metre A m^2
= ampere metre squared (q.v.)

ampere square metre per joule second A m^2/J s
SI unit of gyromagnetic ratio
1 coulomb per kilogramme (q.v.)

ampere-turn (At)
ampère-tour
Amperewindung
depr. and obsol. unit of magnetomotive force
Def: number of turns of a coil times current (in amperes) flowing through the coil

amu; a.m.u. atomic mass unit (old physical)

ångström Å
ångström
Ångström
unit of wavelength
10^{-10} m = **10^{-1}** nm

ap; ap. apothecaries'

apostilb asb
unit of luminance
$3.183\,10 \times 10^{-1}$ cd/m²

apoth; apoth. apothecaries'

apothecaries' drachm, dram, ounce, pound
see: drachm, dram, ounce, pound (apothecaries')

apothecaries' units
Obsolescent units of mass used rarely in UK and US and including the following:

1 apothecaries' ounce (oz apoth in UK, oz ap in US)	= **24** scruples
1 drachm (—) (in UK) 1 dram (dr) (in US)	= **3** scruples
1 scruple (—)	= **20** grains
1 grain (gr) (no symb. in US)	= **1/480** apothecaries' ounce

a.p.s.i. amperes per square inch (A/in^2)

are a
are
Ar
oSI unit of area
10^2 m²
10^{-2} ha
$1.195\,99 \times 10^2$ yd²
Note: used for agrarian measurements only; the hectare (q.v.) is more widely used in practice than the are

aS absiemens

A s ampere-second

asb apostilb

assay ton (UK) (—)
unit of mass
$3.266\,67 \times 10^{-2}$ kg

assay ton (US) (—)
unit of mass
$2.916\,67 \times 10^{-2}$ kg

astronomical unit (AU)
unité astronomique (*UA*)
astronomische Einheit (*AE*)
unit of length
$1.496\,00 \times 10^{11}$ m

at technical atmosphere

At; a.t.; A.T. ampere-turn

ata G obsol. and depr. form of symbol 'at' when used for measuring absolute pressure, cf. psia

At.-Gew. G abbr. for *Atomgewicht* (= atomic weight)

atm standard atmosphere

atmosphere
see: standard atmosphere; technical atmosphere

at. no. atomic number

atomic mass unit (old chemical) (—)
obsol. unit of atomic mass ('weight')
Def: 1 atomic mass unit = 1/16 of the mass of an atom of natural oxygen
1.6601×10^{-27} kg
Note: also called dalton or atomic weight unit

atomic mass unit (old physical) (amu)
obsol. unit of atomic mass ('weight')
Def: 1 amu = 1/16 of the mass of an atom of nuclide ^{16}O
1.6597×10^{-27} kg

atomic mass unit (unified) (u)

unité de masse atomique (unifiée)
atomare Masseneinheit (vereinheitlichte)
unit of (unified) atomic mass constant
Def: 1 u = 1/12 of the mass of an atom of nuclide ^{12}C
$1.660\,44 \times 10^{-27}$ kg
1.000 317 9 amu
1.000 043 atomic mass unit (old chemical)
Note: also unit of (rest) mass of particles, mass excess and mass defect

atomic weight unit (awu)

Note: name sometimes used for atomic mass unit (old chemical) (p.v.)

atta (A)

a prefix denoting $\times 10^{18}$ (not internationally accepted)

atto (a)

a prefix denoting $\times 10^{-18}$. Example: attometre (am)

atu

G obsol. and depr. form of symbol 'at' when used for measuring pressure below atmospheric pressure

atü

G obsol. and depr. form of symbol 'at' when used for measuring gauge pressure, cf. psig

A-turn ampere-turn

at. wt. atomic weight

AU; A.U. astronomical unit

ÅU ångström unit = ångström

aV abvolt

av avoirdupois

avdp avoirdupois

avoirdupois units

Units of mass used in UK and US. They include the following:

1 ton (—)	=	**2240** lb
1 hundredweight (cwt)	=	**112** lb
1 cental (ctl)	=	**100** lb
1 quarter (qr)	=	**28** lb
1 stone (—)	=	**14** lb
1 pound (lb)	=	**0.453 592 37** kg (fundamental)
1 ounce (oz)	=	**1/16** lb
1 dram (dr)	=	**1/16** oz
1 grain (gr)	=	**1/7000** lb

Note: In the US the first two units are called long ton or gross ton, and long hundredweight, respectively, and are rarely used. Instead the following units are used:

1 short ton (—)	=	**2000** lb
1 short hundredweight (—)	=	**100** lb

Cental and stone are not used and grain has no officially approved symbol in the US.

A.W.; Aw

G abbr. for *Amperewindung* = ampere turn

awu atomic weight unit

aΩ abohm

B

b barn

B bel

bar bar

bar

Bar

unit of pressure

$\mathbf{1.0 \times 10^5}$ N/m²

1.0 hpz (hectopièze)

$1.450\ 38 \times 10$ lbf/in²

barn b

barn

Barn

unit of cross section (area)

$= \mathbf{10^{-28}}$ m^2 $= \mathbf{10^2}$ fm^2 (square femtometre)

barn per erg b/erg

unit of spectral cross section

$\mathbf{10^{-21}}$ m^2/J

barn per steardian b/sr

unit of differential cross section

$\mathbf{10^{-28}}$ m^2/sr

barn per steradian erg b/sr erg

unit of spectral differential cross section

$\mathbf{10^{-21}}$ m^2/sr J

barrel (—)

US unit of volume (capacity) for petroleum, etc.

1.589 87 × 10^{-1} m^3

1.589 87 × 10^2 dm^3 or litre

4.2 × **10** gallon (US)

barrel see: dry barrel

barye (ba)

= dyne per square centimetre (q.v.)

baud (—)

unit of telegraph signalling speed

The number of code elements per second. The term code elements includes both pulses and spaces.

Note: sometimes used as special name for the unit bit per second (cf. bit)

bel B

bel

Bel

Note: its sub-multiple decibel (q.v.) is used in practice

BeV billion electronvolts (GeV = 10^9 eV)

B.H.P. British horse-power (hp)

Bi biot

billion

= $\mathbf{10^{12}}$ in European countries, including UK (cf. tera)
= $\mathbf{10^{9}}$ in US (cf. giga)

billionth

= $\mathbf{10^{-12}}$ in European countries, including UK (cf. pico)
= $\mathbf{10^{-9}}$ in US (cf. nano)

biot (Bi)

CGSB unit of electric current
10 A

biot centimetre squared (Bi cm^2)

CGSB unit of electromagnetic moment
10^{-3} A m^2

biot per centimetre (Bi/cm)

CGSB unit of magnetic field strength
7.957 75 × 10 A/m

biot second (Bi s)

CGSB unit of electric charge
10 C

bit (—)

binary unit of:
(a) information content (decision between two equally possible states)
(b) capacity in an information storage device (logarithm to base two of the number of possible states in the device)

bit per unit length (e.g. bit/mm, bit/cm, bit/in)

unit of bit density

bit per unit area (e.g. bit/mm^2, bit/cm^2, bit/in^2)
unit of bit density

Note: the above two units are used as units of information density

bit per second (bit/s)
unit of bit rate
Note: used e.g. as a unit of read-out rate of a storage device. Sometimes called *baud* (q.v.)

Blindwatt (bW, BW)
G depr. unit of reactive power (= var)

blondel (—)
= apostilb (q.v.)

board foot (—)
unit of volume (for timber only)
2.359 74 $\times$ 10^{-3} m^3
1.44 $\times$ **10^2** in^3

bougie nouvelle
name of a unit of luminous intensity changed at 9th CGPM, 1948, to candela (q.v.)

bps bits per second (bit/s)

British thermal unit Btu
unit of heat (energy)
1.055 06 $\times$ 10^3 J
2.519 96 $\times$ 10^{-1} kcal
2.930 71 $\times$ 10^{-4} kWh
7.781 69 $\times$ 10^2 ft lbf
Note: this is the 'international table British thermal unit' adopted by the 5th ICPS, 1956. This unit is used throughout this dictionary

British thermal unit foot per square foot hour degree Fahrenheit
(Btu ft/ft^2 h °F)
unit of thermal conductivity
1.730 73 W/m K

British thermal unit inch per square foot hour degree Fahrenheit
(Btu in/ft^2 h degF)
unit of thermal conductivity
$1.442\,28 \times 10^{-1}$ W/m K
$3.444\,82 \times 10^{-4}$ cal/cm s degC
$1.240\,14 \times 10^{-1}$ kcal/m h degC
$8.333\,33 \times 10^{-2}$ Btu/ft h degF

British thermal unit inch per square foot second degree Fahrenheit (Btu in/ft^2 s °F)
unit of thermal conductivity
$5.192\,20 \times 10^{2}$ W/m K

British thermal unit per cubic foot Btu/ft^3
unit of calorific value (volume basis)
$3.725\,89 \times 10^{4}$ J/m^3
8.899 15 kcal/m^3 or cal/dm^3

British thermal unit per cubic foot degree Fahrenheit
(Btu/ft^3 degF)
unit of thermal capacity per unit volume
$6.706\,61 \times 10^{4}$ J/m^3 K
$1.601\,85 \times 10$ kcal/m^3 degC

British thermal unit per cubic foot hour (Btu/ft^3 h)
unit of heat release
$1.034\,97 \times 10$ W/m^3
8.899 15 kcal/m^3 h

British thermal unit per foot hour degree Fahrenheit
(Btu/ft h degF)
unit of thermal conductivity
1.730 73 W/m K
$4.133\,79 \times 10^{-3}$ cal/cm s degC
1.488 16 kcal/m h degC
1.2×10 Btu in/ft^2 h degF

British thermal unit per foot second degree Fahrenheit
(Btu/ft s degF)
unit of thermal conductivity
$6.230\,64 \times 10^{3}$ W/m K

British thermal unit per hour Btu/h

unit of power

$2.930\ 71 \times 10^{-1}$ W

British thermal unit per pound Btu/lb

unit of specific internal energy and specific latent heat

2.326 $\times$ **10^3** J/kg

$5.555\ 56 \times 10^{-1}$ kcal/kg or cal/g

$2.371\ 86 \times 10^{2}$ kgf m/kg or kpm/kg

$7.781\ 69 \times 10^{2}$ ft lbf/lb

British thermal unit per pound degree Rankine or Fahrenheit

Btu/lb °R or Btu/lb degF

= kilocalorie (I.T.) per kilogramme degree Kelvin (q.v.)

British thermal unit per second foot degree Fahrenheit

(Btu/s ft degF)

= British thermal unit per foot second degree Fahrenheit

British thermal unit per second square foot degree Fahrenheit

(Btu/s ft^2 degF)

= British thermal unit per square foot second degree Fahrenheit (q.v.)

British thermal unit per square foot hour (Btu/ft^2 h)

unit of density of heat flow rate

3.154 59 W/m^2

$7.534\ 61 \times 10^{-5}$ cal/cm^2 s

2.712 46 kcal/m^2 h

$2.930\ 71 \times 10^{-1}$ W/ft^2

$2.035\ 22 \times 10^{-3}$ W/in^2

British thermal unit per square foot hour degree Fahrenheit

(Btu/ft^2 h degF)

unit of coefficient of heat transfer

5.678 26 W/m^2 K

$1.356\ 23 \times 10^{-4}$ cal/cm^2 s degC

4.882 43 kcal/m^2 s degC

British thermal unit per square foot second degree Fahrenheit
(Btu/ft^2 s degF)
unit of coefficient of heat transfer
$2.044\,17 \times 10^4$ W/m^2 K

Btu British thermal unit

B.Th.U.; B.th.u.; BTU British thermal unit (Btu)

bu bushel (US)

bushel (UK) (—)
UK unit of volume (capacity)
$3.636\,87 \times 10^{-2}$ m^3
$3.636\,87 \times 10$ dm^3 or litre
1.284 35 ft^3
1.032 06 bushel (US)
Note: there are also other bushels

bushel (US) (bu)
US unit of volume (capacity) for dry measure
$3.523\,91 \times 10^{-2}$ m^3
$3.523\,91 \times 10$ dm^3 or litre
$2.150\,42 \times 10^3$ in^3 (= definition)
$9.689\,39 \times 10^{-1}$ bushel (UK)

bW; BW G abbr. for *Blindwatt* (q.v.)

C

c centigrade

c centi

c G obsol. symbol for *Kubik* . . .; used only for ccm (cm^3), cdm (dm^3) and cmm (mm^3)

c obsol. symbol for cycle or curie or metric carat

C coulomb; obsol. symbol for curie

°C degree Celsius

cal calorie; here used for calorie I.T.

cal_{15} calorie 15 °C

cal_{IT} calorie I.T.

$cal_{thermochem.}$ thermochemical calorie

Cal obsol. abbr. for kilocalorie (kcal)

calorie see: calorie (I.T.); calorie (15 °C); calorie (thermochemical). When the name calorie is used unspecified after July, 1956, the international table calorie is meant

calorie (15°C) (cal_{15})
calorie 15°C
15 *grd-Kalorie*
unit of heat (energy)
4.185 5 J

calorie (I.T.) cal_{IT} (or cal)
unit of heat (energy)
4.186 8 J
1.163 × **10**$^{-6}$ kWh
3.968 32 × 10^{-3} Btu
Note: cf. international table calorie

calorie (thermochemical) (cal (thermochem.))
US unit of heat (energy)
4.184 0 J
Note: also called defined calorie

calorie (I.T.) per centimetre second degree Celsius
(cal/cm s degC)
unit of thermal conductivity
4.186 8 × **10**^{2} W/m K
3.6 × **10**^{2} kcal/m h degC
2.419 09 × 10^{2} Btu/ft h degF

calorie (I.T.) per cubic centimetre second (cal/cm^3 s)
unit of heat release
4.186 8 × 10^{6} W/m^3

calorie (I.T.) per gramme cal/g
= 1 kilocalorie per kilogramme (q.v.)

calorie (I.T.) per gramme degree Kelvin or Celcius
cal/g °K or cal/g degC
= kilocalorie (I.T.) per kilogramme degree Kelvin (q.v.)

calorie (I.T.) per second centimetre degree (cal/s cm deg)
= calorie (I.T.) per centimetre second degree Celsius (q.v.)

calorie (I.T.) per second square centimetre degree
(cal/s cm^2 deg)
= calorie (I.T.) per square centimetre second degree Celsius (q.v.)

calorie (I.T.) per square centimetre second (cal/cm^2 s)
unit of density of heat flow rate
4.186 8 × $\mathbf{10^4}$ W/m^2
3.6 × $\mathbf{10^4}$ kcal/m^2 h
1.327 21 × 10^4 Btu/ft^2 h

calorie (I.T.) per square centimetre second degree Celsius
(cal/cm^2 s degC)
unit of coefficient of heat transfer
4.186 8 × $\mathbf{10^4}$ W/m^2 deg
3.6 × $\mathbf{10^4}$ kcal/m^2 h degC
7.373 38 × 10^3 Btu/ft^2 h degF

candela cd
candela
Candela
SI base-unit of luminous intensity
Def: The candela is the unit of luminous intensity defined under that name by the CGPM

candela per square centimetre cd/cm^2
unit of luminance
$\mathbf{10^4}$ cd/m^2
Note: also called stilb

candela per square foot cd/ft^2
unit of luminance
$1.076\,39 \times 10$ cd/m^2

candela per square inch cd/in^2
unit of luminance
$1.550\,00 \times 10^3$ cd/m^2

candela per square metre cd/m^2
SI unit of luminance
10^{-4} sb (stilb)
3.141 59 asb or blondel
$3.141\,59 \times 10^{-4}$ L (lambert)
Note: also called nit

carat see: metric carat

cbm G abbr. for *Kubikmeter* = cubic metre (m^3)

$\ldots^{cc}$ one hundredth of a centigrade (centesimal second)

cc; ccm cubic centimetre (cm^3)

cd candela

cent (—)
unit of frequency interval
$8.333\,33 \times 10^{-4}$ octave (= 1/1200)

cental (ctl)
unit of mass
$4.535\,92 \times 10$ kg
10^2 lb

centesimal minute $\ldots^{cg}$
unit of plane angle
10^{-2} grade
Note: name sometimes used for centigrade (q.v.)

centesimal second $\ldots^{cc}$
unit of plane angle
10^{-4} grade
Note: name sometimes used for one hundredth of a centigrade

centi (c)

a prefix denoting $\times 10^{-2}$. It should be avoided as far as possible and used only where well established in practice. Examples: centigrade (. . . .^{cg}), centigramme (cg), centilitre (cl), centimetre (cm), centipoise (cP), centistokes (cSt).

centiare ca

unit of area

1 m^2

Note: should be avoided

centigrade . . . .^{cg}

centigrade

Zentigon; (*Neuminute*)

$\mathbf{10^{-2}}$ grade

Note: also called centesimal minute (q.v.)

centigrade; one hundredth of a ~ . . . .^{cc}

unit of plane angle

$\mathbf{10^{-4}}$ grade

Note: also called centesimal second (q.v.)

Centigrade heat unit (CHU)

unit of energy (enthalpy)

1.8 Btu (q.v.)

centimetre cm

centimètre

Zentimeter

mSI unit and CGS base-unit of length

10^{-2} m

Note: this unit is used as a fundamental unit in many branches of industry, such as building and textile industry

centimetre per second squared cm/s^2

CGS unit of (linear) acceleration

$\mathbf{10^{-2}}$ m/s^2

Note: also called galileo or gal

centimetre second degree Celsius per calorie (I.T.)
(cm s degC/cal)
unit of thermal resistivity
$2.388\,46 \times 10^{-3}$ m K/W
$2.777\,78 \times 10^{-3}$ m h degC/kcal
$4.133\,79 \times 10^{-3}$ ft h degF/Btu

centimetre to the fourth cm^4
CGS unit of second moment of area
10^{-8} metre to the fourth (q.v.)

centipoise cP
unit of (dynamic) viscosity
10^{-2} P
10^{-3} newton second per metre squared (q.v.)

centistokes cSt
unit of kinematic viscosity
10^{-6} metre squared per second (q.v.)

cfm cubic foot per minute (ft^3/min)

cfs cubic foot per second (ft^3/s)

. . . .^{cg} centigon

cg centigramme

CGSe unit
unit of the so-called electrostatic CGS system. See: under stat- (e.g. statampere) or under CGSe unit (e.g. CGSe unit of permittivity)

CGSe unit of magnetic field strength
$2.654\,42 \times 10^{-9}$ A/m

CGSe unit of magnetic flux
$2.997\,92 \times 10^{2}$ Wb

CGSe unit of magnetic flux density
$2.997\,92 \times 10^{6}$ T

CGSe unit of magnetic polarization
$3.767\,30 \times 10^{7}$ T

CGSe unit of magnetization
$2.997\,92 \times 10^{7}$ A/m

CGSe unit of permeability
$1.129\,41 \times 10^{15}$ H/m

CGSe unit of permittivity
$8.854\,19 \times 10^{-12}$ F/m

CGSe unit of reluctance
$8.854\,19 \times 10^{-14}$ H^{-1}

CGSm unit
unit of the so-called electromagnetic CGS system. See: under ab- (e.g. abampere) or under CGSm unit (e.g. CGSm unit of permittivity)

CGSm unit of magnetic polarization
$1.256\,64 \times 10^{-3}$ T

CGSm unit of magnetization
$\mathbf{10^{-3}}$ A/m

CGSm unit of permeability
$1.256\,64 \times 10^{-6}$ H/m

CGSm unit of permittivity
$7.957\,75 \times 10^{9}$ F/m

CGSm unit of reluctance
$7.957\,75 \times 10^{7}$ H^{-1}

ch chain; F abbr. for *cheval vapeur* = metric horsepower (—)

chain (—)
UK and US unit of length
$\mathbf{2.011\,68 \times 10}$ m
22 yd = **66** ft = **792** in
Note: also called Gunter's chain, imperial chain or surveyor's chain. Symbol 'ch' is used in US.

chain see also: engineer's chain

CHU Centigrade heat unit

Ci curie

circular inch (—)
UK and US unit of area
$5.067\,07 \times 10^{-4}$ m^2
$7.853\,98 \times 10^{-1}$ in^2 ($= \pi/4$)
1.0 $\times$ 10^6 circular mils
Note: 1 circular inch = the area of a circle 1 inch in diameter

circular mil (—)
UK and US unit of area
$5.067\,07 \times 10^{-10}$ m^2
$7.853\,98 \times 10^{-7}$ in^2
1.0 $\times$ 10^{-6} circular inch
Note: 1 circular mil = the area of a circle one-thousandth of an inch in diameter

cl centilitre

clusec (—)
unit of leak rate used in vacuum technology
$1.333\,22 \times 10^{-6}$ N m/s
1.0 $\times$ 10^{-2} lusec

cm centimetre

cont hp Continental horse-power
= metric horsepower (—)

cord (—)
unit of volume (for timber only)
3.634 56 m^3
1.28 $\times$ 10^2 ft^3

coulomb C

coulomb

Coulomb

SI unit of electric charge and electric flux

= As

$2.777\,78 \times 10^{-4}$ Ah

For electric charge:

$2.997\,92\ 10^{9}$ sC

$1.000\,00 \times 10^{-1}$ aC

For electric flux:

$3.767\,30 \times 10^{10}$ sC

1.256 64 aC

coulomb metre C m

SI unit of electric dipole moment

= msA

$2.997\,92 \times 10^{11}$ sC cm

$1.000\,00 \times 10$ aC cm

coulomb metre squared per kilogramme $C\ m^2/kg$

SI unit of specific gamma ray constant

$= m^2kg^{-1}s\ A$

coulomb metre squared per volt $C\ m^2/V$

SI unit of polarizability of molecule

$= kg^{-1}s^4A^2$

coulomb per cubic metre C/m^3

SI unit of volume density charge

$= m^{-3}sA$

$2.997\,92 \times 10^{3}$ sC/cm^3

$1.000\,00 \times 10^{-7}$ aC/cm^3

coulomb per kilogramme C/kg

SI unit of ionization exposure

$= kg^{-1}sA = A\ m^2/J\ s$

$3.875\,97 \times 10^{3}$ R (röntgen)

coulomb per kilogramme second C/kg s

SI unit of ionization exposure rate

= A/kg

$3.875\,97 \times 10^{3}$ R/s (röntgen per second)

coulomb per mole C/mol
SI unit of Faraday constant

coulomb per square metre C/m^2
SI unit of surface density of charge, displacement and electric polarization
$= m^{-2}s\ A$
For surface density and polarization:
$2.997\,92 \times 10^5$ sC/cm^2
$1.000\,00 \times 10^{-5}$ aC/cm^2
For displacement:
$3.767\,30 \times 10^6$ sC/cm^2
$1.256\,64 \times 10^{-4}$ aC/cm^2

cP centipoise

cps cycle(s) per second = hertz (Hz)

cSt centistokes

c/s cycle(s) per second = hertz (Hz)

ctl cental

cu; cu.
UK and US abbr. for cubic (e.g. cu ft = ft^3, cu in = in^3, cu yd = yd^3)

cubic centimetre cm^3
centimètre cube
Kubikzentimeter
mSI and CGS unit of volume
10^{-6} m^3 = **1** ml (millilitre)

cubic centimetre per gramme cm^3/g
CGS unit of specific volume
10^{-3} m^3/kg = **1** dm^3/kg or l/kg

cubic centimetre per kilogramme cm^3/kg
unit of specific volume
10^{-6} m^3/kg

cubic decimetre dm^3

decimètre cube

Kubikdezimeter

mSI unit of volume

1.0 × $\mathbf{10^{-3}}$ m^3

1.0 l

2.199 69 × 10^{-1} UKgal

2.641 72 × 10^{-1} USgal

1.759 76 UKpt

2.113 38 liq pt

cubic decimetre per kilogramme dm^3/kg

unit of specific volume

$\mathbf{10^{-3}}$ m^3/kg

1 m^3/t or cm^3/g or l/kg or ml/g

cubic foot ft^3

UK and US unit of volume

2.831 68 × 10^{-2} m^3

2.831 68 × 10 dm^3 or litre

1.726 × $\mathbf{10^3}$ in^3

cubic foot per pound ft^3/lb

FPS unit of specific volume

6.242 80 × 10^{-2} m^3/kg

2.240 × $\mathbf{10^3}$ ft^3/ton

1.728 × $\mathbf{10^3}$ in^3/lb

cubic foot per second ft^3/s

FPS unit of volume rate of flow

2.831 68 × 10^{-2} m^3/s

cubic foot per ton ft^3/ton

unit of specific volume

2.786 96 × 10^{-5} m^3/kg

cubic inch in^3

UK and US unit of volume

1.638 71 × 10^{-5} m^3

1.638 71 × 10 cm^3

cubic inch per pound in^3/lb

unit of specific volume

$3.612\,73 \times 10^{-5}$ m^3/kg

$5.787\,04 \times 10^{-4}$ ft^3/lb

cubic metre m^3

mètre cube

Kubikmeter; *Meter hoch drei*

SI unit of volume

$1.307\,95$	yd^3
$3.531\,47 \times 10$	ft^3
$6.102\,37 \times 10^4$	in^3
$8.107\,1 \times 10^{-4}$	acre foot
$2.199\,69 \times 10^2$	gallon (UK)
$2.641\,72 \times 10^2$	gallon (US)
1.0 $\times$ **10**^{3}	litre (q.v.)

Note: 1 cubic centimetre is not 10^{-2} m^3 but $(10^{-2}\ m)^3 = 10^{-6}\ m^3$

cubic metre per hour m^3/h

unit of volume rate of flow

$2.777\,78 \times 10^{-4}$ m^3/s

$9.809\,63 \times 10^3$ ft^3/s

cubic metre per kilogramme m^3/kg

SI unit of specific volume

1.0 $\times$ **10**^{3} dm^3/kg or l/kg or cm^3/g or ml/g or m^3/t

$1.601\,85 \times 10$ ft^3/lb

$2.767\,99 \times 10^4$ in^3/lb

cubic metre per mole m^3/mol

(SI) unit of molar volume

cubic metre per second m^3/s

SI unit of volume rate of flow

3.6 $\times$ **10**^{3} m^3/h

$3.531\,47 \times 10$ ft^3/s

Note: also unit of volume velocity and recombination coefficient

cubic yard yd^3

UK and US unit of volume

$7.645\,55 \times 10^{-1}$ m^3

2.7 $\times$ **10** ft^3

4.665 6 $\times$ **10^4** in^3

curie Ci

curie

Curie

unit of activity

3.7 × 10^{10} s^{-1}

curie per gramme Ci/g

unit of specific activity

3.7 × 10^{13} s^{-1} kg^{-1}

cusec cubic foot per second (ft^3/s)

CV F abbr. for *cheval vapeur* = metric horsepower (—)

cwt hundredweight

cycle per second c/s

unit of frequency

1 hertz (q.v.)

D

d day; deci

D depr. symbol for deca (da)

da deca

dag decagramme

dalton

Note: name sometimes used for atomic mass unit (old chemical) (q.v.)

daraf (—)

Note: name used in US for reciprocal farad (F^{-1})

day d

jour

Tag

unit of time

8.64 × **10**^{4} s

1.44 × **10**^{3} min

2.4 × **10** h

Note: symbol j is sometimes used in France

deca (da)

a prefix denoting × 10. It should be avoided as far as possible and used only where well established in practice. Examples: decagramme (dag), decajoule (daJ), decalumen (dalm), decanewton (daN).

deci (d)

a prefix denoting × 10^{-1}. It should be avoided as far as possible and used only where well established in practice. Examples: decibel (dB), decigramme (dg), decilitre (dl), decimetre (dm).

decibel dB (or db)

décibel

Dezibel

Note: name used for the pure number 1, which is the unit of ten times the common logarithm ($\log_{10}$) of the ratio of two powers or energies. If the power ratio = the square of an amplitude ratio, 1 dB ≙ 0.115 129 3 Np. Also used as unit of sound power level, sound pressure level and sound reduction index (sound transmission loss)

decimilligrade . . . .^{cc}

depr. unit of plane angle

10^{-4} grade (q.v.)

Note: should be called one hundredth of a centigrade; also called centesimal second (q.v.)

deg

E and F obsol. abbreviation for degree when used as a unit of temperature interval. If unspecified by addition of another letter it refers to degree Kelvin (sometimes abbreviated degK) or degree Celsius (sometimes abbreviated degC), as degK = degC. See: temperature interval

degC

obsol. abbr. for degree Celsius (q.v.) when used as a unit of temperature interval

degF

obsol. abbr. for degree Fahrenheit (q.v.) when used as a unit of temperature interval

degK

obsol. abbr. for degree Kelvin (see: kelvin) when used as a unit of temperature interval

degR

obsol. abbr. for degree Rankine (q.v.) when used as a unit of temperature interval

degree . . . .°

degré

Grad; *Altgrad*

oSI unit of plane angle

1.745 33	$\times 10^{-2}$	rad	($= \pi/180$)
6.0	$\times$ **10**	′ (minute)	
3.6	$\times$ **10^3**	″ (second)	
1.111 11		^{g} (grade)	(= 100/90)
1.111 11	$\times 10^{-2}$	∟ (right angle)	

Note: it can also be subdivided decimally so that e.g. 4° 7′ 30″ is written as 4.125°. The degree with its decimal subdivision is recommended for use when radian is not suitable.

degree (deg)

degré (deg)

Grad (grd)

unit of temperature interval

Note: see deg

degree Celsius °C

degré Celsius

Grad Celsius

unit of Celsius temperature and temperature interval

For temperatures:

$x°\mathrm{C} \mathrel{\hat{=}} (x + 273.15)\ \mathrm{K} \mathrel{\hat{=}} (1.8x + 32)\ °\mathrm{F}$

$\mathrel{\hat{=}} (1.8x + 491.67)\ °\mathrm{R}$

For temperature intervals:

$1°\mathrm{C} = 1\ \mathrm{K} = 1.8$ degF or degR

Note: cf. temperature interval and deg

degree centigrade

depr. name for degree Celsius (q.v.)

degree Fahrenheit °F

degré Fahrenheit

Grad Fahrenheit

unit of Fahrenheit temperature and temperature interval

For temperatures:

$$x°\mathrm{F} \mathrel{\hat{=}} \frac{1}{1.8}(x + 459.67)\ \mathrm{K} \mathrel{\hat{=}} \frac{1}{1.8}(x - 32)\ °\mathrm{C}$$

$$\mathrel{\hat{=}} (x + 459.67)\ °\mathrm{R}$$

For temperature intervals:

$$1°\mathrm{F} = \frac{1}{1.8}\ \mathrm{K}\ \text{or}\ °\mathrm{C} = 1\ \mathrm{degR}$$

degree Kelvin °K

unit of Kelvin temperature and former SI unit of thermodynamic temperature

Note: the name of this unit was changed at the 13th CGPM, 1967, to kelvin (q.v.) and its symbol to K. Also cf. temperature interval and deg

degree per second °/s

unit of angular velocity

$1.745\,33 \times 10^{-2}$ rad/s

1.047 20 rad/min

degree per second squared $^\circ/s^2$

unit of angular acceleration

$1.745\,33 \times 10^{-2}$ rad/s²

degree Rankine °R

degré Rankine

Grad Rankine

unit of Rankine temperature and temperature interval

For temperatures:

$$x\,^\circ\mathrm{R} \triangleq \frac{1}{1.8}\,x\,\mathrm{K} \triangleq \frac{1}{1.8}(x - 491.67) \triangleq x - 459.67$$

For temperature intervals:

$$1\,^\circ\mathrm{R} = \frac{1}{1.8}\,\mathrm{K} \text{ or } ^\circ\mathrm{C} = 1 \text{ degF}$$

degree Réaumur (°R)

obsol. and depr. unit of temperature

Both for temperatures and temperature intervals:

x °R (degree Réaumur) = $0.8x$ °C

Note: °R is used in this dictionary for degree Rankine (q.v.)

dg decigramme

dipotre (dpt)

dioptrie

Dioptrie

unit of power of a lens

1 m^{-1}

dk depr. symbol for deca (da)

dkg depr. symbol for decagramme (dag)

dl decilitre

dm decimetre

Doppelzentner (dz)

G obsol. unit of mass

100 kg = **1** quintal (q.v.)

dpt; dptr dioptre

dr dram (avoirdupois)

dr.ap.; dr ap apothecaries' drachm (or dram)

drachm, apothecaries' (—)

UK unit of mass

$3.887\,93 \times 10^{-3}$ kg

1.25 $\times$ $\mathbf{10^{-1}}$ oz (apoth)

6.0 $\times$ **10** gr

Note: the same as dram (apothecaries')

dram, apothecaries' (dr ap)

US unit of mass

= drachm (apothecaries') (q.v.)

dram, avoirdupois (dr)

UK and US unit of mass

$1.771\,85 \times 10^{-3}$ kg

$3.906\,25 \times 10^{-3}$ lb

6.25 $\times$ $\mathbf{10^{-2}}$ oz

$2.734\,38 \times 10$ gr (grain)

dry barrel (bbl)

US unit of volume (capacity) for dry measure

$1.156\,27 \times 10^{-1}$ m^3

$1.156\,27 \times 10^{2}$ dm^3 or litre

7.056 $\times$ $\mathbf{10^{3}}$ in^3

Note: this is the standard barrel used for fruits, vegetables and dry commodities

dry pint (dry pt)

US unit of volume (capacity) for dry measure

$5.506\,10 \times 10^{-4}$ m^3

$5.506\,10 \times 10^{-1}$ dm^3 or litre

9.689×10^{-1} UKpt

dry quart (US) (dry qt)

US unit of volume (capacity) for dry measure

$1.101\,221$ dm^3

dt G abbr. for *Dezitonne* = quintal (q.v.)

dwt pennyweight

dyn dyne

dyne dyn
dyne
Dyn
CGS unit of force
= g cm/s^2
10^{-5} N

dyne centimetre dyn cm
CGS unit of moment of force
10^{-7} N m = **1** erg (q.v.)

dyne centimetre per biot (dyn cm/Bi)
CGSB unit of magnetic flux
10^{-8} Wb

dyne centimetre second dyn cm/s
CGS unit of moment of momentum
10^{-7} kg m^2/s = **1** g cm^2/s = erg second (q.v.)

dyne per biot centimetre (dyn/Bi cm)
CGSB unit of magnetic flux density and magnetic polarization
For magnetic flux density:
10^{-4} T
For magnetic polarization:
1.256 64 × 10^{-3} T

dyne per biot squared (dyn/Bi^2)
CGSB unit of (non-rationalized) permeability
1.256 64 × 10^{-6} H/m

dyne per centimetre dyn/cm
CGS unit of surface tension
= erg/cm^2
10^{-3} N/m

dyne per cubic centimetre dyn/cm^3
CGS unit of specific weight
10 N/m^3

dyne per franklin (dyn/Fr)
CGSF unit of electric field strength
2.997 92 × 10^4 V/m

dyne per square centimetre dyn/cm^2
CGS unit of pressure
10^{-1} N/m^2 = **10**$^{-6}$ bar = **1** μbar
Note: also called barye

dyne second dyn/s
CGS unit of momentum
10^{-5} kg m/s = **1** g cm/s

dyne second per centimetre dyn s/cm
CGS unit of mechanical impedance
10^{-3} N s/m

dyne second per centimetre cubed dyn s/cm^3
CGS unit of specific acoustic impedance
10 N s/m^3

dyne second per centimetre to the fifth dyn s/cm^5
CGS unit of acoustic impedance
10^5 N s/m^5

dz: G abbr. for *Doppelzentner*

E

electronvolt eV
électronvolt
Elektronvolt; *Elektronenvolt*
unit of energy
= 1 e × 1 V (e = elementary charge (q.v.))
1.602 10 × 10^{-19} J (C × V = J)

electronvolt per centimetre eV/cm
unit of linear stopping power and linear energy transfer
1.602 10 × 10^{-17} J/m

electronvolt per square centimetre second $eV/cm^2\ s$

unit of energy flux density

$1.602\ 10 \times 10^{-15}\ W/m^2$

electronvolt square centimetre $eV\ cm^2$

unit of atomic stopping power

$1.602\ 10 \times 10^{-23}\ J\ m^2$

electronvolt square centimetre per gramme $eV\ cm^2/g$

unit of mass stopping power

$1.602\ 10 \times 10^{-20}\ J\ m^2/kg$

emE

G abbr. for *elektromagnetische Einheit* = electromagnetic unit

emu; e.m.u.

abbr. used for CGSm units (e.g. emu of current, emu of permittivity). See: under ab- (e.g. abampere) or under CGSm unit (e.g. CGSm unit of permittivity)

engineer's chain (—)

unit of length

3.048 $\times$ **10** m

1.0 $\times$ **10^2** ft

erg erg

erg

Erg

CGS unit of work (energy)

10^{-7} joule (q.v.) = **1** dyn cm

erg per biot squared (erg/Bi^2)

CGSB unit of self inductance and mutual inductance

10^{-9} H

erg per biot squared (erg/Bi^2)

CGSB unit of self inductance and mutual inductance

10^{-9} H

erg per centimetre erg/cm

CGS unit of linear stopping power and linear energy transfer

10^{-5} J/m

1 dyn

erg per cubic centimetre erg/cm^3

CGS unit of energy density and calorific value (volume basis)

10^{-1} J/m^3

erg per cubic centimetre degree Celsius erg/cm^3 degC

CGS unit of thermal capacity per unit volume

10^{-1} J/m^3 K

erg per cubic centimetre nanometre erg/cm^3 nm

CGS unit of spectral radiant energy density

10^8 J/m^4

erg per cubic centimetre second erg/cm^3 s

CGS unit of heat release

10^{-1} W/m^3

erg per centimetre second degree Celsius erg/cm s degC

CGS unit of thermal conductivity

10^{-5} W/m K

erg per franklin (erg/Fr)

CGSF unit of electric potential

$2.997\,92 \times 10^2$ V

erg per gramme erg/g

CGS unit of specific internal energy and specific latent heat

10^{-4} J/kg

erg per gramme degree Celsius erg/g degC

CGS unit of specific heat capacity

10^{-4} J/kg K

erg per gramme second erg/g s
CGS unit of absorbed dose rate and kerma rate
10^{-4} W/kg
10^{-2} rad/s

erg per second erg/s
CGS unit of power
10^{-7} W

erg per second steradian erg/s sr
CGS unit of radiance
10^{-7} W/sr

erg per second steradian square centimetre erg/s sr cm^2
CGS unit of radiance
10^{-3} W/sr m^2

erg per square centimetre erg/cm^2
= dyne per centimetre (q.v.)

erg per square centimetre second erg/cm^2 s
CGS unit of density of heat flow rate
10^{-3} W/m^2

erg per square centimetre second degree Celsius
erg/cm^2 s degC
CGS unit of coefficient of heat transfer
10^{-3} W/m^2 K

erg per square centimetre second kelvin to the fourth power
erg/cm^2 s K^4
CGS unit of Stefan-Boltzmann constant
10^{-3} W/m^2 K^4

erg second erg s
CGS unit of action
= dyn cm s
10^{-7} J s or N m s

erg square centimetre erg cm^2
CGS unit of atomic stopping power
10^{-11} J m^2

erg square centimetre per gramme erg cm^2/g

CGS unit of mass stopping power

10^{-8} J m^2/kg

erg square centimetre per second erg cm^2/s

CGS unit of first radiation constant

10^{-11} W m^2

esE

G abbr. for *elektrostatische Einheit* = electrostatic unit (esu)

esu; e.s.u.

abbr. used for CGSe units (e.g. esu of current, esu of permittivity). See: under stat- (e.g. statampere) or under CGSe unit (e.g. CGSe unit of permittivity)

eV electronvolt

F

f femto

F farad; femta

°F degree Fahrenheit

farad F

farad

Farad

SI unit of capacitance

$= m^{-2}\ kg^{-1}\ s^4\ A^2 = C/V = As/V = s/\Omega$

$8.987\ 55 \times 10^{11}$ sF

$1.000\ 00 \times 10^{-9}$ aF

farad per metre F/m

SI unit of permittivity

$= m^{-3}\ kg^{-1}\ s^4\ A^2$

$1.129\ 41 \times 10^{11}$ CGSe unit

$1.256\ 64 \times 10^{-10}$ CGSm unit

fathom (—)
unit of length
1.828 8 m = **2** yd
Note: for marine use

fbm board foot

fc footcandle

femta (F)
a prefix denoting $\times 10^{15}$ (not internationally accepted)

femto (f)
a prefix denoting $\times 10^{-15}$. Example: femtometre (fm)

fermi (—)
unit of length
$\mathbf{10^{-15}}$ m = **1** fm (femtometre)
Note: used for nuclear distances

fg frigorie

fl dr fluid drachm (UK); fluid dram (US)

fl oz fluid ounce

fluid drachm (UK) (UK fl dr)
UK unit of volume (capacity)
$3.551\,63 \times 10^{-6}$ m^3

fluid dram (US) (US fl dr)
US obsol. unit of volume (capacity) for liquid measure
$3.696\,69 \times 10^{-3}$ m^3

fluid ounce (UK) (UK fl oz)
UK unit of volume (capacity)
$2.841\,31 \times 10^{-5}$ m^3
$2.841\,31 \times 10$ cm^3
1.733 87 in^3
$9.607\,60 \times 10^{-1}$ US fl oz

fluid ounce (US) see: liquid ounce

fm fathom (—)

foot ft

UK and US unit, and FPS and ft-lbf-s base-unit of length

3.048 $\times$ **10**$^{-1}$ m

1.893 94 $\times$ 10^{-4} mile

3.333 33 $\times$ 10^{-1} yd

1.2 $\times$ **10** in

1.645 79 $\times$ 10^{-4} nautical mile (internat.)

1.644 74 $\times$ 10^{-4} nautical mile (UK)

foot see also: survey foot

foot cubed ft^3

unit of modulus of section

Note: the same as cubic foot (q.v.)

foot hour degree Fahrenheit per British thermal unit

(ft h degF/Btu)

unit of thermal resistivity

5.777 89 $\times$ 10^{-1} m K/W

2.419 09 $\times$ 10^{2} cm s degC/cal

6.719 69 $\times$ 10^{-1} m h degC/kcal

foot of water (conventional) (ftH_2O)

unit of pressure

2.989 07 $\times$ 10^{3} N/m^2

foot per minute ft/min

unit of velocity

5.08 $\times$ **10**$^{-3}$ m/s

3.048 $\times$ **10**$^{-1}$ m/min

foot per second ft/s

FPS unit of velocity

3.048 $\times$ **10**$^{-1}$ m/s

1.097 28 km/h

foot per second squared ft/s^2

FPS unit of acceleration

3.048 00 $\times$ **10**$^{-1}$ m/s^2

foot pound-force (ft lbf)

ft-lbf-s unit of work

1.355 82 J

3.766 16 × 10^{-7} kWh

foot pound-force per pound (ft lbf/lb)

unit of specific internal energy and specific latent heat

2.989 07 J/kg

7.139 26 × 10^{-4} kcal/kg

3.048 × 10^{-1} kgf m/kg or kpm/kg

1.285 07 × 10^{-3} Btu/lb

foot pound-force per pound degree Fahrenheit

(ft lbf/lb degF)

unit of specific heat capacity

5.380 32 J/kg K

1.285 07 × 10^{-3} kcal/kg degC or Btu/lb degF

5.486 4 × 10^{-1} kgf m/kg degC or kpm/kg degC

foot pound-force per second (ft lbf/s)

ft-lbf-s unit of power

1.355 82 W

foot poundal ft pdl

FPS unit of work

4.214 01 × 10^{-2} J

1.750 56 × 10^{-8} kWh

foot poundal per second ft pdl/s

FPS unit of power

4.214 01 × 10^{-2} W

foot squared per hour ft^2/h

unit of kinematic viscosity

2.580 64 × 10^{-5} m^2/s

2.580 64 × 10 cSt

2.777 78 × 10^{-4} ft^2/s

foot squared per second ft^2/s

FPS unit of kinematic viscosity

$9.290\,30 \times 10^{-2}$ m^2/s

$9.290\,30 \times 10^{4}$ cSt

3.6 $\times$ $\mathbf{10^3}$ ft^2/h

foot to the fourth ft^4

FPS unit of second moment of area

$8.630\,97 \times 10^{-3}$ m^4

foot-candle (fc)

unit of illumination

$1.076\,39 \times 10$ lx or lm/m^2

foot-lambert (ft L)

unit of luminance

$3.426\,26$ cd/m^2

fr frigorie

Fr franklin

franklin (Fr)

CGSF unit of electric charge and electric flux

For electric charge:

$3.335\,64 \times 10^{-10}$ C

For electric flux:

$2.654\,42 \times 10^{-11}$ C

franklin centimetre (Fr cm)

CGSF unit of electric dipole moment

$3.335\,64 \times 10^{-12}$ C m

franklin per second (Fr/s)

CGSF unit of electric current

$3.335\,64 \times 10^{-10}$ A

franklin per square centimetre (Fr/cm^2)

CGSF unit of electric polarization and displacement

For polarization:

$3.335\,64 \times 10^{-6}$ C/m^2

For displacement:

$2.654\,42 \times 10^{-7}$ C/m^2

franklin squared per erg (Fr^2/erg)
CGSF unit of capacitance
$1.112\,65 \times 10^{-12}$ F

franklin squared per erg centimetre ($Fr^2/erg\ cm$)
CGSF unit of permittivity
$8.854\,19 \times 10^{-12}$ F/m

frigorie (fg)
unit of heat (for refrigeration)
$4.185\,50 \times 10^3$ J
1.0 $kcal_{15}$

ft foot

ftH_2O conventional foot of water

ft L; ft la foot lambert

furlong (—)
US and UK unit of length
$\mathbf{2.011\,68 \times 10^2}$ m = $\mathbf{2.2 \times 10^2}$yd

G

. . . .^{g} gon; grade

g gramme

G giga; gauss

g; G
unit of acceleration equal to standard gravity (q.v.)

gal Gal
gal
Gal
unit of (linear) acceleration
$\mathbf{10^{-2}}$ m/s^2 = **1** cm/s^2
Note: also called galileo

gal gallon

Gal gal

galileo see: gal

gallon (UK) (UKgal)

UK unit of volume (capacity)

$4.546\,09 \times 10^{-3}$ m^3

4.546 09 dm^3 or litre

$1.605\,44 \times 10^{-1}$ ft^3

$2.774\,20 \times 10^{2}$ in^3

1.200 95 USgal

Note: also called imperial gallon. It is the fundamental UK unit of capacity defined by the WMA, 1963, unfortunately with reference to densities in gramme per (old) millilitre. Due to redefinition of litre (q.v.) in 1964 the factor for conversion of gallons to litres changed

gallon (UK) per hour (UKgal/h)

UK unit of volume rate of flow

$1.262\,83 \times 10^{-6}$ m^3/s

$4.546\,09 \times 10^{-3}$ m^3/h

gallon (UK) per mile (UKgal/mile)

UK unit of fuel consumption

2.825 litres/km

gallon (UK) per minute (UKgal/min)

UK unit of volume rate of flow

$7.576\,81 \times 10^{-7}$ m^3/s

$2.727\,65 \times 10^{-1}$ m^3/h

gallon (UK) per pound (UKgal/lb)

UK unit of specific volume

$1.002\,24 \times 10^{-2}$ m^3/kg

gallon (UK) per second (UKgal/s)

UK unit of volume rate of flow

$4.546\,09 \times 10^{-3}$ m^3/s

$1.636\,59 \times 10$ m^3/h

gallon (US) (USgal)

US unit of volume (capacity) for liquid measure

$3.785\,41 \times 10^{-3}$ m^3

3.785 41 dm^3 or litre

$1.336\,81 \times 10^{-1}$ ft^3

2.31 $\times$ **10**^{2} in^3 (= definition)

$8.326\,74 \times 10^{-1}$ UKgal

gallon (US) per mile (USgal/mile)

US unit of fuel consumption

2.352 litres/km

gamma (γ)

unit of mass

10^{-9} kg = **1** μg

gauss (Gs)

CGSm unit of magnetic flux density

10^{-4} T

1 dyn/(Bi cm)

Note: the name gauss, abbreviated G, was also used for a unit of rationalized magnetic flux density (magnetic induction) G = 10^{-4} V s/m^2 or Wb/m^2 or T

Gb gilbert

gee pound see: slug

gf gramme-force

gi gill (US)

giga (G)

a prefix denoting $\times 10^9$. Examples: gigaelectronvolt (GeV), gigahertz (GHz), gigajoule (GJ), giganewton (GN), gigapond (Gp), gigawatt (GW).

gilbert (Gb)

CGSm unit of magnetomotive force

$7.957\,75 \times 10^{-1}$ A

gilbert per centimetre (Gb/cm)
CGSm unit of magnetic field strength
Note: corresponds to oersted (q.v.)

gilbert per maxwell (Gb/Mx)
CGSm unit of reluctance
$7.957\,75 \times 10^{7}\ H^{-1}$

gill (UK) (—)
UK unit of volume (capacity)
$1.420\,65 \times 10^{-4}\ m^{3}$

gill (US) (gi)
US unit of volume (capacity) for liquid measure
$1.182\,94 \times 10^{-4}\ m^{3}$

gm gramme (g)

gon $\ldots .^{g}$
unit of plane angle
Note: see grade

gr grain

g rad gramme rad

grade $\ldots .^{g}$
grade
Gon; *Neugrad*
oSI unit of plane angle
$1.570\,80 \times 10^{-2}$ rad (= $\pi/200$)
0.9° (degree) = **54′** (minute) = **3240″** (second)
0.01 ∟ (right angle)
Note: also called gon; symbol (gr) sometimes used in France. The unit grade with its decimal subdivision is recommended for use when radian is not suitable

grade per second ^{g}/s
unit of angular velocity
$1.570\,80 \times 10^{-2}$ rad/s

grade per second squared $^g/s^2$

unit of angular acceleration

$1.570\,80 \times 10^{-2}$ rad/s^2

grain (gr)

UK and US unit of mass

6.479 891 $\times$ $\mathbf{10^{-5}}$ kg

6.479 891 $\times$ **10** mg

$1.428\,57 \times 10^{-4}$ lb (= 1/7000)

Note: no symbol is used in US

grain per hundred cubic feet (gr/100 ft^3)

unit of concentration (mass density)

$2.288\,35 \times 10^{-5}$ kg/m^3

grain per UK gallon (gr/UKgal)

UK unit of concentration (mass density)

$1.425\,38 \times 10^{-2}$ kg/m^3

grain per US gallon (gr/USgal)

US unit of concentration (mass density)

$1.711\,81 \times 10^{-2}$ kg/m^3

gramme g

gramme

Gramm

mSI unit and CGS base-unit of mass

$\mathbf{10^{-3}}$ kg

gramme centimetre per second g cm/s

CGS unit of momentum

$\mathbf{10^{-5}}$ kg m/s = **1** dyn s

gramme centimetre per second squared g cm/s^2

= dyne (q.v.)

gramme centimetre squared g cm^2

CGS unit of moment of inertia

$\mathbf{10^{-7}}$ kg m^2

gramme centimetre squared per second g cm^2/s
CGS unit of moment of momentum
10^{-7} kg m^2/s = **1** dyn cm/s

gramme per cubic centimetre g/cm^3
CGS unit of (mass) density
10^3 kg/m^3
1 kg/dm^3 (q.v.)

gramme per litre g/l
unit of (mass) density
1 kg/m^3

gramme per millilitre g/ml
unit of (mass) density
10^3 kg/m^3
1 kg/dm^3 (q.v.)

gramme per square metre g/m^2
10^{-3} kilogramme per square metre (q.v.)

gramme rad g rad
unit of energy imparted
10^{-5} J
10^2 erg
Note: also known as gram-rad

gram-rad see: gramme-rad

grd G obsol. abbr. for *Grad* = degree

Gs gauss

H

h hecto; hour

H henry

ha hectare

hand (—)

obsol. unit of length
1.016×10^{-1} m = 4 in

hectare ha

hectare
Hektar
oSI unit of area
$\mathbf{10^4}$ m = $\mathbf{10^2}$ a = $\mathbf{10^{-2}}$ km^2
$3.861\,02 \times 10^{-3}$ mile^2
2.471 05 acre
9.884 22 rood
$1.195\,99 \times 10^4$ yd^2
Note: used for agrarian measurements only

hectare-millimetre (ha × mm)

unit of volume
10 m^3
Note: used very rarely

hectolitre hl

hectolitre
Hektoliter
unit of volume (capacity)
$\mathbf{10^{-1}}$ m^3 = $\mathbf{10^2}$ dm^3 or litre
Note: used e.g. in brewing industry

hecto (h)

a prefix denoting $\times 10^2$. It should be avoided as far as possible and used only where well established in practice. Examples: hectare (ha), hectolitre (hl), hectopieze (hpz)

hectopièze (hpz)

unit of pressure
$\mathbf{10^5}$ N/m^2 = **1** bar
Note: used widely in France

Hefner candle

obsol. unit of luminous intensity
0.92 cd

henry H

henry

Henry

SI unit of self inductance, mutual inductance and permeance

$= m^2 kg\, s^{-2} A^{-2} = Vs/A = Wb/A = \Omega s$

For inductances:

$1.112\,65 \times 10^{-12}$ sH

$1.000\,00 \times 10^{9}$ aH

For permeance:

$8.854\,19 \times 10^{-14}$ sH

$7.957\,75 \times 10^{7}$ aH

henry per metre H/m

SI unit of permeability

$= m\, kg/s^2 A^2 = Wb/A\, m = Vs/A\, m$

$8.854\,19 \times 10^{-16}$ CGSe unit

$7.957\,75 \times 10^{5}$ CGSm unit

hertz Hz

hertz

Hertz

SI unit of frequency

1 s^{-1} = **1** c/s

hl hectolitre

horsepower (metric) (—)

cheval-vapeur (ch)

Pferdestärke (PS)

unit of power

$7.354\,99 \times 10^{2}$ W

7.5 × **10** kgf m/s or kp m/s

$9.863\,20 \times 10^{-1}$ hp

horsepower (hp)

UK and US unit of power

$7.457\,00 \times 10^{2}$ W

5.5 × **10²** ft lbf/s

1.013 87 metric horsepower

horsepower hour (metric) (—)
cheval-heure (ch h)
Pferdestärkenstunde (PSh)
unit of work
2.6478×10^6 J

horsepower hour hp h
UK and US unit of energy
$2.684\,52 \times 10^6$ J
$7.457\,00 \times 10^{-1}$ kWh

hour h
heure
Stunde
unit of time
$\mathbf{3.6} \times \mathbf{10^3}$ s = $\mathbf{6.0} \times \mathbf{10}$ min
$4.166\,67 \times 10^{-2}$ d

hp horsepower

hp h horsepower hour

hpz hectopièze

hundredweight (cwt)
UK unit of mass
$5.080\,23 \times 10$ kg
$\mathbf{1.12} \times \mathbf{10^2}$ lb
1.12 short hundredweight
Note: also called long hundredweight to distinguish it from short hundredweight (q.v.)

hundredweight see also: short hundredweight

hyl (—)
m-kgf-s unit of mass
= gf s^2/m or p s^2/m
$\mathbf{9.806\,65} \times \mathbf{10^{-3}}$ kg
Note: rarely used

Hz hertz

i.; imp. imperial

ipm inches per minute (in/min)

ipr inches per revolution (in/rev)

imperial (imp.)
Adjective indicating that a particular unit is one of the so-called imperial units (q.v.)

imperial units
Imperial units lawful for use for trade in the United Kingdom under the Weight and Measures Act, 1963 are the following:
(a) units of length: mile, furlong, chain, yard, foot, inch
(b) units of area: square mile, acre, rood, square yard, square foot, square inch
(c) units of volume: cubic yard, cubit foot, cubic inch
(d) units of capacity: gallon, quart, pint, gill, fluid ounce, (fluid drachm, minim)
(e) units of mass or 'weight': ton, hundredweight, cental, quarter, stone, pound, ounce, dram, grain, ounce troy, (ounce apothecaries, drachm, scruple)
Note: ounce troy may be used only for articles made of gold, silver or other precious metals, and the units in parentheses, which should be avoided, only for drugs

in inch

inHg conventional inch of mercury

inH_2O conventional inch of water

inch in
(*pouce*)
(*Zoll*)
UK and US unit of length
2.54 $\times$ **10^{-2}** m
2.777 78 $\times$ 10^{-2} yd
8.333 33 $\times$ 10^{-2} ft

inch cubed in^3

unit of modulus of section

Note: the same as cubic inch (q.v.)

inch of mercury (conventional) (inHg)

unit of pressure

$3.386\,39 \times 10^3$ N/m^2

inch of water (conventional) (inH_2O)

unit of pressure

$2.490\,89 \times 10^2$ N/m^2

inch per minute in/min

unit of velocity

$4.233\,33 \times 10^{-4}$ m/s

2.54 $\times$ **10**$^{-2}$ m/min

inch per second in/s

unit of velocity

2.54 $\times$ **10**$^{-2}$ m/s

inch squared per hour in^2/h

unit of kinematic viscosity

$1.792\,11 \times 10^{-7}$ m^2/s

$1.792\,11 \times 10^{-1}$ cSt

6.451 6 $\times$ **10**$^{-4}$ m^2/h

$1.929\,01 \times 10^{-6}$ ft^2/s

inch squared per second in^2/s

unit of kinematic viscosity

6.451 6 $\times$ 10^{-4} m^2/s

6.451 6 $\times$ **10**^{2} cSt

$6.944\,44 \times 10^{-3}$ ft^2/s

inch to the fourth in^4

unit of second moment of area

$4\,162\,31 \times 10^{-7}$ m^4

INM

abbr. sometimes used for international nautical mile

international ampere (A_{int})
unit of electric current
$9.998\,5 \times 10^{-1}$ A

international candle (—)
obsol. unit of luminous intensity
1.02 cd

international coulomb (C_{int})
unit of electric charge
$9.998\,5 \times 10^{-1}$ C

international farad (F_{int})
unit of capacitance
$9.995\,1 \times 10^{-1}$ F

international henry (H_{int})
unit of inductance and permeance
1.000 49 H

international joule (mean) (J_{int})
unit of work, energy and heat
1.000 19 J
Note: obsolete from 1948

international nautical mile see: nautical mile (international)

international ohm (Ω_{int})
unit of resistance
1.000 49 Ω

international table calorie cal_{IT}
calorie I.T.
internationale Tafel-Kalorie
unit of heat (energy)
Note: in this dictionary referred to as calorie (I.T.) and the symbol cal is used for it, as no other calorie should be used internationally

international table kilocalorie $kcal_{IT}$

kilocalorie I.T.

Internationale Tafel-Kilokalorie

unit of heat (energy)

Note: in this dictionary referred to as kilocalorie (I.T.) and the symbol kcal is used for it, as no other kilocalorie should be used internationally

international volt (V_{int})

unit of electric potential

1.000 34 V

international watt (W_{int})

unit of power

1.000 19 W

J

J joule

joule J

joule

Joule

SI unit of work, energy and heat

$= m^2\ kg/s^2 = N\ m = W\ s = C \times V$

$2.777\ 78 \times 10^{-7}$ kWh

$1.019\ 72 \times 10^{-1}$ kgf m

$2.388\ 46 \times 10^{-4}$ kcal

$7.375\ 62 \times 10^{-1}$ ft lbf

$9.478\ 17 \times 10^{-4}$ Btu

Note: also unit of enthalpy, Gibbs function and level width

joule per cubic metre J/m^3

SI unit of electromagnetic energy density

$= kg/m\ s^2 = W\ s/m^2 = N/m^2$

1.0 × **10** erg/cm^3

$2.388\ 46 \times 10^{-4}\ kcal/m^3$

$2.683\ 92 \times 10^{-5}\ Btu/ft^3$

Note: also unit of radiant energy density, sound energy density and calorific value (volume basis)

joule per cubic metre kelvin J/m^3 K

SI unit of thermal capacity per unit volume

$= m^{-1}\ kg\ s^{-2}\ K^{-1}$

2.388 46 × 10^{-4} kcal/m^3 degC

1.0 × 10 erg/cm^3 degC

1.491 07 × 10^{-5} Btu/ft^3 degF

joule per degree Celsius (J/degC)

= joule per kelvin (q.v.)

joule per kelvin J/K

SI unit of heat capacity and entropy

$= m^2\ kg\ s^{-2}\ K^{-1}$ = J/degC

2.388 46 × 10^{-4} kcal/degC

Note: also unit of Boltzmann constant

joule per kilogramme J/kg

SI unit of specific internal energy and specific latent heat

$= m^2/s^2$ = N m/kg

2.388 46 × 10^{-4} kcal/kg or cal/g

1.0 × 10^4 erg/g

1.019 72 × 10^{-1} kgf m/kg or kpm/kg

4.299 23 × 10^{-4} Btu/lb

3.345 53 × 10^{-1} ft lbf/lb

Note: also unit of calorific value (mass basis), absorbed dose and kerma

joule per kilogramme degree Celsius J/kg °C or J/kg degC

= joule per kilogramme kelvin (q.v.)

joule per kilogramme degree Kelvin J/kg °K or J/kg degK

= joule per kilogramme kelvin (q.v.)

joule per kilogramme kelvin J/kg K

SI unit of specific heat capacity and specific entropy

$= m^2 s^{-2} K^{-1}$ = J/kg degC

2.388 46 × 10^{-4} kcal/kg degC or Btu/lb degF

1.0 × 10^4 erg/g degC

1.019 72 × 10^{-1} kgf m/kg degC or kpm/kg degC

1.858 63 × 10^{-1} ft lbf/lb degF

joule per metre J/m

SI unit of linear stopping power and linear energy transfer

= m kg s^{-2} = newton (q.v.)

10^5 erg/cm or dyne

joule per metre to the fourth power J/m^4

SI unit of spectral (concentration of) radiant energy density (in terms of wavelength)

= m^{-2} kg s^{-2}

10^{-8} erg/cm^3 nm

joule per mole J/mol

(SI) unit of molar internal energy

= m^2 kg s^{-2} mol^{-1}

Note: also unit of chemical potential and affinity

joule per mole degree Celsius J/(mol °C)

= joule per mole kelvin (q.v.)

joule per mole kelvin J/mol K

(SI) unit of molar heat capacity and molar entropy

= m^2kg $s^{-2}K^{-1}mol^{-1}$

Note: also unit of molar gas constant

joule per pound degree Kelvin or Celsius

(J/lb °K or J/lb degC)

unit of specific heat capacity (mass basis) and specific entropy

2.204 62 J/kg K

joule per second J/s

= watt (q.v.)

joule per square metre J/m^2

SI unit of impact strength

= kg/s^2

= newton per metre (q.v.)

joule second J s

SI unit of action

= m^2 kg/s = N m s

10^7 erg cm or dyn cm s

Note: also unit of Planck constant

joule square metre J m^2

SI unit of atomic stopping power

= m^4 kg s^{-2}

10^{11} erg cm^2

joule square metre per kilogramme J m^2/kg

SI unit of mass stopping power

= m^4 s^{-2}

10^8 erg cm^2/g

Julian year (—)

unit of time

3.155 76 × **10^7** s

5.259 6 × **10^5** min

8.766 × **10^3** h

3.652 5 × **10^2** d

K

k kilo

K kelvin

°K degree Kelvin

kc kilocycle (per second) = kilohertz (kHz)

kcal kilocalorie

$kcal_{15}$ kilocalorie 15 °C

$kcal_{IT}$ kilocalorie I.T.

$kcal_{thermochem.}$ thermochemical calorie

kelvin K

kelvin

Kelvin

SI base-unit of thermodynamic temperature and SI unit of temperature interval

Def: The kelvin is the unit of thermodynamic temperature defined under that name by the CGPM

For temperatures:

x K $\triangleq$ $(x - 273.15)$ °C $\triangleq$ $(1.8x - 459.67)$ °F
$\triangleq$ $1.8x$ °R

For temperature intervals:

1 K = 1°C = 1.8 degF or degR

kg kilogramme

kgf kilogramme-force

kgph kilogramme per hour (kg/h)

kgpm kilogramme per minute (kg/min)

kgps kilogramme per second (kg/s)

kilo (k)

a prefix denoting $\times 10^3$. Examples: kiloampere (kA), kilobar (kbar), kilocalorie (kcal), kilocoulomb (kC), kiloelectronvolt (keV), kilogramme (kg), kilohertz (kHz), kilojoule (kJ), kilometre (km), kilomole (kmol), kilonewton (kN), kiloohm (kΩ), kilopond (kp), kilosecond (ks), kilovolt (kV), kilowatt (kW)

kilocalorie (I.T.) $kcal_{IT}$ (or kcal)

unit of heat (energy)

10^3 I.T. calorie (q.v.)

kilocalorie (I.T.) metre per square metre hour degree Celsius

(kcal m/m^2 h °C)

unit of thermal conductivity

1.163 W/m K

kilocalorie (I.T.) per cubic metre kcal/m^3

unit of calorific value (volume basis)

4.186 8 $\times$ **10^3** J/m^3

1.123 70 $\times$ 10^{-1} Btu/ft^3

kilocalorie (I.T.) per cubic metre degree Celsius

(kcal/m^3 degC)

unit of thermal capacity per unit volume

4.186 8 $\times$ **10^3** J/m^3 K

6.242 80 $\times$ 10^{-2} Btu/ft^3 degF

kilocalorie (I.T.) per cubic metre hour (kcal/m^3 h)

unit of heat release

1.163 W/m^3

1.123 70 $\times$ 10^{-1} Btu/ft^3 h

kilocalorie (I.T.) per degree Celsius (kcal/degC)

unit of heat capacity

4.186 8 J/K

kilocalorie (I.T.) per hour kcal/h

unit of power

1.163 W

kilocalorie (I.T.) per kilogramme kcal/kg

unit of specific internal energy and specific latent heat

4.186 8 $\times$ **10^3** J/kg

1.0 cal/g

4.269 35 $\times$ 10^2 kgf m/kg or kpm/kg

1.8 Btu/lb

1.400 70 $\times$ 10^3 ft lbf/lb

kilocalorie (I.T.) per kilogramme degree Celsius

(kcal/kg degC)

unit of specific heat capacity

4.186 8 $\times$ **10^3** J/kg K

1.0 Btu/lb degF

7.781 69 $\times$ 10^2 ft lbf/lb degF

kilocalorie (I.T.) per metre hour degree Celsius
(kcal/m h degC)
unit of thermal conductivity
1.163 W/m K
2.777 78 × 10^{-3} cal/cm s degC
6.719 69 × 10^{-1} Btu/ft h degF

kilocalorie (I.T.) per square metre hour (kcal/m^2 h)
unit of density of heat flow rate
1.163 W/m^2
2.777 78 × 10^{-5} cal/cm^2 s
3.686 69 × 10^{-1} Btu/ft^2 h

kilocalorie (I.T.) per square metre hour degree Celsius
(kcal/m^2 h degC)
unit of coefficient of heat transfer
1.163 W/m^2 K
2.777 78 × 10^{-5} cal/cm^2 s degC
2.048 16 × 10^{-1} Btu/ft^2 h degF

kiloelectronvolt per micrometre keV/μm
1.602 10 × 10^{-10} J/m

kilogramme kg
kilogramme
Kilogramm
SI base-unit of mass
Def: The kilogramme is the unit of length defined under that name by the CGPM
10^{-3} t = **10^{-2}** q
5.0 × **10^3** metric carat
2.204 62 lb
3.527 40 × 10 oz
1.543 24 × 10^4 gr

kilogramme metre per second kg m/s
SI unit of momentum
1.0 × **10^5** g cm/s
7.233 01 lb ft/s

kilogramme metre per second squared kg m/s^2
= newton (q.v.)

kilogramme metre squared $kg\ m^2$

SI unit of moment of inertia

1.0 $\times$ **10^7** $g\ cm^2$

2.373 04 $\times$ 10 $lb\ ft^2$

3.417 17 $\times$ 10^3 $lb\ in^2$

kilogramme metre squared per second $kg\ m^2/s$

SI unit of moment of momentum

1.0 $\times$ **10^7** $g\ cm^2/s$

2.373 04 $\times$ 10 $lb\ ft^2/s$

kilogramme per cubic centimetre kg/cm^3

unit of (mass) density

10^6 kg/m^3

kilogramme per cubic decimetre kg/dm^3

unit of (mass) density

10^3 kg/m^3 = **1** kg/l or g/cm^3 or g/ml or t/m^3 or Mg/m^3

kilogramme per cubic metre kg/m^3

SI unit of (mass) density

1.0 $\times$ **10^{-3}** g/cm^3

6.242 80 $\times$ 10^{-2} lb/ft^3

4.369 96 $\times$ 10^4 $gr/100\ ft^3$

Note: also unit of mass concentration

kilogramme per hectare kg/ha

unit of mass per unit area

10^{-4} kilogramme per square metre (q.v.)

kilogramme per hour kg/h

unit of mass rate of flow

2.777 78 $\times$ 10^{-4} kg/s

6.123 95 $\times$ 10^{-4} lb/s

2.204 62 lb/h

9.842 07 $\times$ 10^{-4} ton/h

kilogramme per litre kg/l

unit of (mass) density

= kilogramme per cubic decimetre (q.v.)

kilogramme per metre kg/m

SI unit of mass per unit length

1.0 t/km

$6.719\,69 \times 10^{-1}$ lb/ft

$5.599\,74 \times 10^{-2}$ lb/in

2.015 91 lb/yd

$8.999\,58 \times 10^{-1}$ ton/1000 yd

Note: used for wires etc.

kilogramme per metre second kg/m s

= newton second per metre squared (q.v.)

kilogramme per mole kg/mol

(SI) unit of molar mass

10^3 g/mol or kg/kmol

kilogramme per second kg/s

SI unit of mass rate of flow

3.6×10^3 kg/h

2.204 62 lb/s

$7.936\,64 \times 10^3$ lb/h

3.543 14 ton/h

kilogramme per square metre kg/m^2

SI unit of mass per unit area

$2.048\,16 \times 10^2$ $lb/1000\ ft^2$

$8.921\,79 \times 10^3$ lb/acre

3.277 06 oz/ft^2

$2.949\,35 \times 10$ oz/yd^2

$2.549\,08 \times 10^3$ $ton/mile^2$

Note: used in agriculture and for sheet metal and plating. Also unit of mean mass range.

kilogramme-force (kgf)

kilogramme-force (kgf)

Kilopond (kp)

m-kgf-s base-unit of force

9.806 65 N

$7.093\,16 \times 10$ pdl

2.204 62 lbf

kilogramme-force metre (kgf m)

mètre-kilogramme-force

Kilopondmeter

m-kgf-s unit of moment of force

9.806 65 N m or J

7.233 01 lbf ft

kilogramme-force metre per kilogramme (kgf m/kg)

m-kgf-s unit of specific internal energy and specific latent heat

9.806 65 J/kg

kilogramme-force metre per kilogramme degree Celsius (kgf/kg degC)

m-kgf-s unit of specific heat capacity

9.806 65 J/kg K

kilogramme-force metre per second (kgf m/s)

m-kgf-s unit of power

9.086 65 W

kilogramme-force metre second (kgf m s)

m-kgf-s unit of action

9.806 65 J s or N m s

kilogramme-force metre second squared (kgf m s^2)

m-kgf-s unit of moment of inertia

9.806 65 kg m^2

kilogramme-force per centimetre (kgf/cm)

m-kgf-s unit of surface tension

9.806 65 $\times$ 10^2 N/m

kilogramme-force per cubic metre (kgf/m^3)

m-kgf-s unit of specific weight

9.806 65 N/m^3

kilogramme-force per metre (kgf/m)

m-kgf-s unit of surface tension

9.806 65 N/m

kilogramme-force per metre second degree Celsius
(kgf/m s degC)
m-kgf-s unit of coefficient of heat transfer
9.806 65 W/m^2 K

kilogramme-force per second degree Celsius (kgf/s degC)
m-kgf-s unit of thermal conductivity
9.806 65 W/m K

kilogramme-force per square centimetre (kgf/cm^2)
m-kgf-s unit of pressure
9.806 65 $\times$ **10^4** N/m^2
9.806 65 $\times$ **10^{-1}** bar
1.0 at
1.422 33 $\times$ 10 lbf/in^2

kilogramme-force per square metre (kgf/m^2)
m-kgf-s unit of pressure
9.806 65 N/m^2
1.0 $\times$ **10^{-4}** kgf/cm^2 (q.v.)

kilogramme-force second (kgf s)
m-kgf-s unit of momentum
9.806 65 kg m/s

kilogramme-force second per square metre (kgf s/m^2)
m-kgf-s unit of (dynamic) viscosity
9.806 65 N s/m^2

kilogramme-force second squared per metre to the fourth
(kgf s^2/m^4)
m-kgf-s unit of density
9.806 65 kg/m^3

kilohyl (khyl)
m-kgf-s unit of mass
= metric technical unit of mass (q.v.)

kilohyl per cubic metre ($khyl/m^3$)
unit of density
Note: name sometimes used for kilogramme-force second squared per metre to the fourth (q.v.)

kilometre km

mSI unit of length

1.0 $\times$ **10^3** m

6.213 71 $\times 10^{-1}$ mile

kilometre per hour km/h

unit of velocity

2.777 78 $\times 10^{-1}$ m/s

6.213 71 $\times 10^{-1}$ mile/h

kilomole kmol

10^3 mole (q.v.)

kilopond (kp)

m-kp-s unit of force

= kilogramme-force (q.v.)

Note: name used in some countries instead of kilogramme-force. In all entries in this dictionary 'kilopond' may be substituted for 'kilogramme-force' to change m-kgf-s units to m-kp-s units. Kilopond was widely used particularly in Germany (from 1939)

kilowatt kW

mSI unit of power

10^3 W

kilowatt hour kWh

kilowattheure

Kilowattstunde

unit of energy

3.6 $\times$ **10^6** J

8.598 45 $\times 10^2$ kcal

3.412 14 $\times 10^4$ Btu

3.670 98 $\times 10^5$ kgf m

km kilometre

kmc; KMC kilomegacycles (per second)

= gigahertz (GHz)

kn knot (international)

knot (international) kn

nœud

Knoten

unit of velocity

= n mile/h

$5.144\,44 \times 10^{-1}$ m/s

1.852 km/h

1.150 78 mile/h

$9.993\,61 \times 10^{-1}$ UK knot

knot (UK) (—)

unit of velocity

$5.147\,73 \times 10^{-1}$ m/s

1.853 18 km/h

1.000 64 kn (internat. knot)

1.0 UK nautical mile per hour

kp kilopond

kph kilometre per hour (km/h)

kW kilowatt

kWh kilowatt hour

L

l litre

L lambert

La; la lambert

lambert (La)

unit of luminance

$3.183\,10 \times 10^{3}$ cd/m^{2}

l.atm litre atmosphere

Lb pound-force (lbf)

lb pound

lbf pound-force

lb tr troy pound

li link (—)

light year (l.y.)
année de lumière (a.l.)
Lichtjahr (Lj)
unit of length
$9.460\,5 \times 10^{15}$ m

link (—)
unit of length
$\mathbf{2.011\,68 \times 10^{-1}}$ m
7.92 in
1.0 $\times$ $\mathbf{10^{-2}}$ chain
Note: also called Gunter's or surveyor's link

liq oz liquid ounce

liq pt liquid pint

liq qt liquid quart

liquid ounce (US) (US liq oz)
US unit of volume (capacity) for liquid measure
$2.957\,35 \times 10^{-5}$ m^3
$2.957\,35 \times 10^{-2}$ dm^3 or litre
1.804 69 in^3
1.040 84 UK fl oz
Note: also called fluid ounce

liquid pint (US) (US liq pt)
US unit of volume (capacity) for liquid measure
$4.731\,76 \times 10^{-4}$ m^3
$4.731\,76 \times 10^{-1}$ dm^3 or litre
$1.671\,01 \times 10^{-2}$ ft^3
$\mathbf{2.887\,5 \times 10}$ in^3
$8.326\,74 \times 10^{-1}$ UKpt

liquid quart (US) (US liq qt)
US unit of volume (capacity) for liquid measure
$9.463\,53 \times 10^{-4}$ m^3

litre l

litre

Liter

oSI unit of volume (capacity)

2.200×10^{-1} gallon (UK)

2.642×10^{-1} gallon (US)

1.760 pint (UK)

2.113 liquid pint (US)

Note: the litre was defined by XII CGPM, 1964, as 1 dm^3 exactly, but its use for precision measurements was discouraged; therefore the conversion factors above have only 4 significant places (cf. cubic decimetre). US spelling of this unit is liter

litre (old) l

depr. unit of volume (capacity)

$1.000\,028 \times 10^{-3}$ m^3

Note: used 1901–1964

litre atmosphere l.atm

unit of work (energy)

$1.013\,28 \times 10^{2}$ J

litre per 100 kilometres l/100 km

unit of fuel consumption

$\frac{282.5}{\text{litres}}$ mile/UKgal

$\frac{235.2}{\text{litres}}$ mile/USgal

litre per kilogramme l/kg

unit of specific volume

= cubic decimetre per kilogramme (q.v.)

litre per mole (l/mol)

unit of molar volume

10^{-3} m^3/mol

Lj G abbr. for *Lichtjahr* = light year

lm lumen

lumen lm

lumen

Lumen

SI unit of luminous flux

= cd sr

lumen hour lm h

unit of quantity of light

3.6 × 10³ lm s

lumen per square foot lm/ft^2

= foot-candle (q.v.)

lumen per square metre lm/m^2

SI unit of luminous exitance

= cd sr/m^2

= lux (q.v.)

lumen per watt lm/W

SI unit of luminous efficacy

= $m^{-2}kg^{-1}s^3$cd sr

lumen second lm s

SI unit of quantity of light

= cd sr s

lusec (—)

unit of leak rate used in vacuum technology

$1.333\,22 \times 10^{-4}$ N m/s

lux lx

lux

Lux

SI unit of illumination

= cd sr/m^2

= lumen per square metre (q.v.)

lux hour lx h

unit of light exposure

3.6 × 10³ lx s

lux second lx s

SI unit of light exposure

= cd sr s m^{-2}

lx lux

l.y. light year

M

m metre; milli

M mega; maxwell

ma myria

Mach number (Mach)

Mach number is the ratio (v/c) of the velocity (v) of a moving object to the velocity of sound (c) in the same medium and under the same conditions. It is used to express the velocity of aircraft. Mach 1 = the velocity of sound, Mach 2 = twice the velocity of sound etc. Velocity of sound in dry air at 0°C is about 331.46 m/s.

magnetic ohm

Note: name sometimes used for gilbert per maxwell (q.v.)

maxwell Mx

CGSm unit of magnetic flux

10^{-8} Wb

1 dyn cm/Bi

Note: The name maxwell, abbreviated M, was also used for a unit of rationalized magnetic flux, M = 10^{-8} Vs or Wb

maxwell per square centimetre (Mx/cm^2)

= gauss (q.v.)

mb millibar

mbar millibar

mc; Mc megacycle (per second) = megahertz (MHz)

MCM thousand circular mils (cf. circular mil)

mega (M)

a prefix denoting $\times 10^6$. Examples: megaampere (MA), megacalorie (Mcal), megadyne (Mdyn), megaelectronvolt (MeV), megaerg (Merg), megagramme (Mg), megahertz (MHz), megajoule (MJ), meganewton (MN), megapond (Mp), megavolt (MV), megawatt (MW), megohm (MΩ)

megagramme Mg

mSI unit of mass

10^3 kg = 1 t

Note: rarely used; tonne is used in practice

megapond (Mp)

10^3 kilopond (q.v.)

metre m

mètre

Meter

SI base-unit of length

Def: The metre is the unit of length defined under that name by the CGPM

$5.399\,57 \times 10^{-4}$	nautical mile (international)
$5.396\,12 \times 10^{-4}$	nautical mile (UK)
$6.213\,71 \times 10^{-4}$	mile
1.093 61	yd
3.280 84	ft
$3.937\,01 \times 10$	in

Note: US spelling of this unit is meter. Also unit of breadth, height, thickness, radius, diameter, length of path, wavelength, sound particle displacement, mean free path, stopping equivalent, mean linear range, and diffusion coefficient for neutron flux density

metre cubed m^3

mètre cube

Kubikmeter; (*Meter hoch drei*)

SI unit of modulus of section

Note: the same as cubic metre (q.v.)

metre hour degree Celsius per kilocalorie (I.T.)

(m h degC/kcal)

unit of thermal resistivity

8.598 45	$\times 10^{-1}$	m K/W
3.6	$\times$ **10^2**	cm s degC/cal
1.488 16		ft h degF/Btu

metre kelvin m K

SI unit for second radiation constant

metre kelvin per watt m K/W

SI unit of thermal resistivity

$= m^{-1} kg^{-1} s^3 K = J^{-1} m s K$

4.186 8	$\times$ **10^2**	cm s degC/cal
1.163		m h degC/kcal
1.730 73		ft h degF/Btu

metre per second m/s

SI unit of velocity

3.6		km/h
6.0	$\times$ **10**	m/min
3.280 84		ft/s
2.236 94		mile/h
1.943 84		kn (international knot)
1.942 60		UK knot

metre per second cubed m/s^3

SI unit of jerk

Note: the conversion factors for m/s and m/s^2 are applicable

metre per second squared m/s^2

SI unit of acceleration

10^2 gal $\approx$ 3.280 84 ft/s^2

metre squared m^2

SI unit of slowing down area, diffusion area and migration area

Note: the same as square metre (q.v.)

metre squared per hour m^2/h

unit of kinematic viscosity

$2.777\ 78 \times 10^{-4}$ m^2/s

$2.777\ 78 \times 10^{2}$ cSt

metre squared per newton second $m^2/N\ s$

= square metre per newton second (q.v.)

metre squared per second m^2/s

SI unit of kinematic viscosity

1.0 $\times$ **10^4** St (stokes) or cm^2/s

1.0 $\times$ **10^6** cSt

3.6 $\times$ **10^3** m^2/h

$1.076\ 39 \times 10$ ft^2/s

$3.875\ 01 \times 10^4$ ft^2/h

$1.550\ 00 \times 10^3$ in^2/s

$5.580\ 01 \times 10^6$ in^2/h

Note: also unit of thermal diffusivity, diffusion coefficient and thermal diffusion coefficient

metre to the fourth m^4

SI unit of second moment of area

10^8 cm^4 = 10^{12} mm^4

$1.158\ 62 \times 10^2$ ft^4

$2.402\ 51 \times 10^6$ in^4

metric carat (—)

carat métrique

metrisches Karat

unit of mass

2×10^{-4} kg

2×10^{2} mg

Note: adopted by 4th CGPM, 1907, for diamonds, fine pearls and precious stones

metric technical unit of mass (—)

unité technique métrique de masse
technische Masseneinheit (TME or ME)
m-kgf-s unit of mass
1 kgf s^2/m or kp s^2/m
9.806 65 kg
2.162 00 lb
6.719 69 $\times$ 10^{-1} slug
Note: sometimes called kilohyl

mev megaelectronvolt (MeV)

mf millifarad (mF); microfarad (μF)

mF millifarad

mfd microfarad (μF)

. . . .^{mg} milligrade

mg milligramme

mGal; mgal milligal (mGal)

mho

name sometimes used in practice for reciprocal ohm (q.v.)

mi US abbr. for mile. Consequently mi^2 = square mile and mi^3 = cubic mile

micro (μ)

a prefix denoting $\times$ 10^{-6}. Examples: microampere (μA), microbar (μbar), microcoulomb (μC), microcurie (μCi), microfarad (μF), microgramme (μg), microhenry (μH), microinch (μin), microjoule (μJ), micrometre (μm), microohm ($\mu\Omega$), microsecond (μs), microvolt (μV), microwatt (μW)

microbar μbar

= dyne per square centimetre (q.v.)

micro-inch μin
unit of length
2.54 × 10^{-8} metre
1.0 × 10^{-6} inch

micrometre μm
micromètre
Mikrometer
mSI unit of length
10^{-6} m
Note: also called micron

micron (μ)
micron
Mikron
Note: more correctly called micrometre (q.v.)

mil (—)
Note: short for milli-inch (q.v.)

mile mile
UK and US unit of length
1.609 344 × 10^3 m
1.76 × 10^3 yd
5.28 × 10^3 ft
Note: also known as statute mile. Abbr. mi is used in the USA.

mile
see also: nautical mile (international) and nautical mile (UK)

mile per gallon (UK) mile/UKgal
unit of reciprocal fuel consumption
0.354 km/litre
$\frac{282.5}{\text{miles}}$ litres/100 km
0.833 mile/USgal

mile per gallon (US) mile/USgal

unit of reciprocal fuel consumption

0.425 km/litre

$\frac{235.2}{\text{miles}}$ litres/100 km

1.20 mile/UKgal

mile per hour mile/h

UK and US unit of velocity

4.470 4 × **10**$^{-1}$ m/s

1.609 344 km/h

1.466 67 ft/s

milli (m)

a prefix denoting × 10^{-3}. Examples: milliampere (mA), millibar (mbar), millicurie (mCi), milligal (mgal), milligrade (. . . .^{mg}), milligramme (mg), millihenry (mH), milli-inch (min), millijoule (mJ), millilitre (ml), millimetre (mm), millinewton (mN), milliohm (mΩ), millipieze (mpz), millipond (mp), milliradian (mrad), millisecond (ms), millitesla (mT), millivolt (mV), milliwatt (mW)

millibar mbar (or mb)

unit of pressure

10^{-3} bar (q.v.)

Note: used in meterological barometry

milligal mGal

unit of acceleration

10^{-5} m/s^2 = **10**$^{-3}$ Gal

milligrade . . . .^{mg}

milligrade

Milligon

unit of plane angle

10^{-3} grade (q.v.)

milligramme per litre ml/l

unit of concentration (mass density)

10^{-3} kg/m^3

milli-inch (min)

unit of length

2.54 × 10^{-5} m

1.0 × 10^{-3} in

Note: sometimes called mil

millilitre ml

millilitre

Milliliter

unit of volume (capacity)

10^{-6} m^3

1 cm^3

10^{-3} litre

millimetre mm

millimètre

Millimeter

mSI unit of length

10^{-3} m

Note: used as fundamental unit in engineering

millimetre of mercury (conventional) (mmHg)

millimètre de mercure conventionnel

Millimeter Quecksilbersäule

unit of pressure

1.333 22 × 10^2 N/m^2

1.333 22 × 10^{-3} bar

1.000 00 torr

millimetre of water (conventional) (mmH_2O)

millimètre d'eau conventionnel

Millimeter H_2O (cf. mmWS)

unit of pressure

9.806 65 N/m^2

9.806 65 × 10^{-5} bar

1.0 kgf/m^2 or kp/m^2

millimicron mμ see: nanometre

min minute; minim (UK or US); milli-inch

minim (UK) (UKmin)

UK unit of volume (capacity)

$5.919\,39 \times 10^{-8}$ m^3

$5.919\,39 \times 10$ mm^3

$3.612\,23 \times 10^{-3}$ in^3

$9.607\,6 \times 10^{-1}$ USmin

minim (US) (USmin)

US unit of volume (capacity) for liquid measure

$6.161\,15 \times 10^{-8}$ m^3

$6.161\,15 \times 10$ mm^3

1.040 84 UKmin

minute min

minute

Minute

unit of time

6.0 $\times$ **10** s

$1.666\,67 \times 10^{-2}$ h

$6.944\,44 \times 10^{-4}$ d

Note: symbol (mn) is sometimes used in France

minute . . . .′

minute

Minute; *Altminute*

oSI unit of plane angle

$2.908\,88 \times 10^{-4}$ rad

$1.666\,67 \times 10^{-2}$ ° (degree)

6.0 $\times$ **10^2** ″ (second)

$1.851\,85 \times 10^{-2}$ ᵍ (grade)

$1.851\,85 \times 10^{-4}$ ∟ (right angle)

minute see: centesimal minute

ml millilitre

mm millimetre

mmHg conventional millimetre of mercury

mmH_2O conventional millimetre of water

mmQS

G abbr. for *Millimeter Quecksilbersäule* = conventional millimetre of mercury (mmHg)

mmWS

G abbr. for *Millimeter Wassersäule* = (non-conventional) millimetre of water

mol mole

mole mol

mole

Mol

(SI) unit of amount of substance

Def: 1 mol is an amount of substance of a system which contains as many elementary units as there are carbon atoms in 0.012 kg (exactly) of ^{12}C. The elementary unit must be specified and may be an atom, a molecule, an ion, an electron, etc. or a specified group of such particles.

Note: recommended by IUPAP, IUPAC and ISO/TC 12

mole per cubic metre mol/m^3

(SI) unit of molarity

mole per kilogramme mol/kg

(SI) unit of molality

mole per litre mol/l

unit of molarity

$\mathbf{10^3}$ mol/m^3

m.p.g. miles per gallon (mile/gal)

m.p.h. miles per hour (mile/h)

ms millisecond

mu micro (μ)

mu, Mu micron = micrometre (μm)

Mx maxwell

myria (ma)
depr. prefix denoting $\times 10^4$

N

n nano

N newton; neper (Np)

nano (n)
a prefix denoting $\times 10^{-9}$. Examples: nanofarad (nF), nanohenry (nH), nanometre (nm)

nanometre nm
mSI unit of length
$\mathbf{10^{-9}}$ m
Note: used to be called millimicron

nautical mile (international) (n mile)
mille marin; *mille* (—)
Seemeile (sm)
oSI unit of length
1.852 $\times$ $\mathbf{10^3}$ m
1.852 km
9.993 61 $\times 10^{-1}$ nautical mile (UK)

nautical mile (UK) (—)
UK unit of length
1.853 18 $\times 10^3$ m
1.853 18 km
1.000 64 n mile (international)
6.08 $\times$ $\mathbf{10^3}$ ft

neper Np (or N)
néper
Neper
Note: name used for the pure number 1, which is the unit of natural logarithm ($\log_e$) of the ratio of two amplitudes. If the amplitude ratio = square root of a power ratio, 1 Np $\triangleq$ 8.685 890 dB. Also used as unit of logarithmic decrement

neper per metre Np/m (or N/m)
unit of attenuation coefficient
1 m^{-1}

neper per second Np/s (or N/s)
unit of damping coefficient
1 s^{-1}

newton N
newton
Newton
SI unit of force
= kg m/s^2
1.0 × $\mathbf{10^{5}}$ dyn
1.0 × $\mathbf{10^{-3}}$ sn
1.019 72 × 10^{-1} kgf or kp
7.233 01 pdl
2.248 09 × 10^{-1} lbf
1.003 61 × 10^{-4} tonf
3.596 94 ozf

newton metre N m
mètre-newton
Newtonmeter
SI unit of moment of force
= m^2 kg/s^2 = J
1.0 × $\mathbf{10^{7}}$ dyn cm
1.019 72 × 10^{-1} kgf m or kp m
2.373 04 × 10 pdl ft
7.375 62 × 10^{-1} lbf ft
8.850 75 lbf in
3.292 69 × 10^{-4} tonf ft
1.416 12 × 10^{2} ozf in

newton metre second N m s
unit of moment of momentum
= kg m^2/s

newton per cubic metre N/m^3

SI unit of specific weight (q.v.)

$= m^{-2}\ kg\ s^{-2}$

10^{-1} dyn/cm^3

newton per metre N/m

SI unit of surface tension

$= kg/s^2 = J/m^2$

1.0 × **10^3** dyn/cm or erg/cm^2

1.019 72 × 10^{-1} kgf/m or kp/m

Note: also unit of spring stiffness rate and force per unit length

newton per square metre N/m^2

SI unit of pressure and stress

$= m^{-1}\ kg\ s^{-2} = J/m^3$

1.0 × **10^{-5}** bar

1.0 × **10** dyn/cm^2 or μbar

1.019 72 × 10^{-1} kgf/m^2 or kp/m^2

1.019 72 × 10^{-5} at or kgf/cm^2 or kp/cm^2

9.869 23 × 10^{-6} atm

7.500 62 × 10^{-3} torr

6.719 69 × 10^{-1} pdl/ft^2

2.088 54 × 10^{-2} lbf/ft^2

1.450 38 × 10^{-4} lbf/in^2

9.323 85 × 10^{-6} $tonf/ft^2$

6.474 90 × 10^{-8} $tonf/in^2$

Note: N/m^2 is in some countries called *pascal* (q.v.); also unit of Young's modulus, shear modulus, bulk modulus and fugacity

newton second N s

unit of momentum

$= kg\ m/s$

newton second per metre N s/m

SI unit of mechanical impedance

$= kg\ s^{-1}$

1.0 × **10^3** dyn s/m

newton second per metre cubed $N\ s/m^3$

SI unit of specific acoustic impedance

$= m^{-2}\ kg\ s^{-1}$

10^{-1} dyn s/cm^3

newton second per metre squared $N\ s/m^2$

SI unit of (dynamic) viscosity

= kg/m s

1.0 × **10** P (poise)

$1.019\ 72 \times 10^{-1}$ kgf s/m^2 or kp s/m^2

$6.719\ 69 \times 10^{-1}$ pdl s/ft^2 or lb/ft s

$2.088\ 54 \times 10^{-2}$ lbf s/ft^2 or slug/ft s

$5.801\ 51 \times 10^{-6}$ lbf h/ft^2

Note: called *poiseuille* (Pl) in France

newton second per metre to the fifth $N\ s\ /m^5$

SI unit of acoustic impedance

$= m^{-4}\ kg\ s^{-1}$

10^{-5} dyn s/cm^5

newton second per square metre $N\ s/m^5$

= newton second per metre squared (q.v.)

newton square metre per ampere $N\ m^2/A$

SI unit of magnetic dipole moment

= weber metre (q.v.)

nit (nt)

unit of luminance

= candela per square metre (q.v.)

Nm^3 1 m^3 of gas under 'standard' conditions

(= 0°C and 1 atm)

Note: used in Germany and some other Continental countries

n mile international nautical mile

nox (nx)

obsol. unit of scotopic illumination

10^{-3} lx

Np neper

nt nit

nx nox

O

O octave (—)

octave (—)

octave
Oktave
unit of frequency interval
The octave is the interval between any two frequencies having a ratio of 2 to 1.

Oe oersted

oersted (Oe)

CGSm unit of magnetic field strength
$7.957\,75 \times 10$ A/m
Note: the name oersted was also used for a unit of rationalized magnetic field strength, Oe = $7.957\,75 \times 10$ A/m

ohm Ω

ohm
Ohm
SI unit of resistance, impedance, modulus of impedance, reactance
$= m^2\ kg\ s^{-3}\ A^{-2} = V/A = W/A^2$
$1.112\,65 \times 10^{-12}$ sΩ
$1.000\,00 \times 10^{9}$ aΩ

ohm metre Ω m

SI unit of resistivity
$1.112\,65 \times 10^{-10}$ sΩ cm
$1.000\,00 \times 10^{11}$ aΩ cm

ohm second Ω s

unit of self inductance
= henry (q.v.)
Note: also called secohm

ohm square millimetre per metre $\Omega\ mm^2/m$
unit of resistivity
10^{-6} Ω m

ounce (oz)
UK and US unit of mass
$2.834\ 95 \times 10^{-2}$ kg
$2.834\ 95 \times 10$ g
6.25 $\times$ **10^{-2}** lb
4.375 $\times$ **10^{2}** gr
Note: this is an avoirdupois unit

ounce, apothecaries' (oz apoth; oz ap)
UK and US unit of mass
= **1** ounce, troy (q.v.)
Note: the abbreviation oz apoth is used in UK, and oz ap in US

ounce, avoirdupois see: ounce

ounce, fluid see: fluid ounce

ounce, liquid see: liquid ounce

ounce, troy (oz tr)
UK and US unit of mass
$3.110\ 35 \times 10^{-2}$ kg
1.0 oz apoth
4.8 $\times$ **10^{2}** gr

ounce inch squared (oz in^2)
unit of moment of inertia
$1.829\ 00 \times 10^{-5}$ kg m^2
$4.340\ 28 \times 10^{-4}$ lb ft^2

ounce per gallon (UK) (oz/UKgal)
unit of concentration (mass density)
6.236 03 kg/m^3

ounce per gallon (US) (oz/USgal)
unit of concentration (mass density)
7.489 15 kg/m^3

ounce per inch (oz/in)

unit of mass per unit length

1.116 12 kg/m

ounce per square foot (oz/ft^2)

unit of mass per unit area

3.051 52	$\times 10^{-1}$	kg/m^2
3.051 52	$\times 10^{2}$	g/m^2
9.0		oz/yd^2
6.25	$\times$ **10**	lb/1000 ft^2
2.722 5	$\times$ **10^{3}**	lb/acre
7.778 57	$\times 10^{2}$	$ton/mile^2$

ounce per square yard (oz/yd^2)

unit of mass per unit area

3.390 57	$\times 10^{-2}$	kg/m^2
3.390 57	$\times 10$	g/m^2
1.111 11	$\times 10^{-1}$	oz/ft^2
6.944 44		lb/1000 ft^2
3.025	$\times$ **10^{2}**	lb/acre
8.642 86	$\times 10$	$ton/mile^2$

ounce-force (ozf)

unit of force

2.780 14	$\times 10^{-1}$	N
2.010 88		pdl
6.25	$\times$ **10^{-2}**	lbf

ounce-force inch (ozf in)

unit of moment of force

7.061 55	$\times 10^{-3}$	N m
7.200 78	$\times 10^{-4}$	kgf m or kp m
5.208 33	$\times 10^{-3}$	lbf ft

oz ounce (avoirdupois)

oz ap; oz apoth apothecaries' ounce

ozf ounce-force

oz t; oz tr troy ounce

p pico; pond

P poise

Pa pascal

parsec (pc)

parsec (pc)
Parsek (Pk)
unit of length
3.086 $\times 10^{16}$ m
3.260 light year
2.062 645 × 10⁵ astronomical unit

pascal (Pa)

unit of pressure
= newton per square metre (q.v.)
Note: used in some countries; in France officially used from 1961

pc parsec; per cent (%)

pdl poundal

peck (—)

UK unit of volume (capacity)
$9.092\ 18 \times 10^{-3}$ m^3

peck (pk)

US unit of volume (capacity) for dry measure
$8.809\ 77 \times 10^{-3}$ m^3

pennyweight (dwt)

unit of mass
$1.555\ 17 \times 10^{-3}$ kg
2.4 × 10 gr
Note: this is a Troy unit

perch (—)

depr. unit of length
= rod (q.v.)

ph phot

phon (—)
phone
Phon
(SI) unit of loudness level

phot (ph)
unit of illumination
10^4 lm/m^2 = **1** lm/cm^2

phot-second (ph s)
unit of light exposure
10^4 lx s

pico (p)
a prefix denoting $\times 10^{-12}$. Examples: picofarad (pF), picometre (pm)

pièze (pz)
MTS unit of pressure
10^3 N/m^2

pint (UK) (UKpt)
UK unit of volume (capacity)
$5.682\,61 \times 10^{-4}$ m^3
$5.682\,61 \times 10^{-1}$ dm^3 or litre
$2.006\,79 \times 10^{-2}$ ft^3
$3.467\,74 \times 10$ in^3
1.200 95 US liq pt

pint see: dry pint, liquid pint

pk US peck

Pk G abbr. for *Parsek* = parsec

Pl poiseuille

poise P
poise
Poise
CGS unit of (dynamic) viscosity
= dyn s/cm^2 = g/cm s
10^{-1} newton second per metre squared (q.v.)

poiseuille Pl

unit of (dynamic) viscosity
= newton second per metre squared (q.v.)
Note: used in France

pole (—)

depr. unit of length
= rod (q.v.)

poncelet (—)

F obsol. unit of power
9.806 65 × 10^2 W

pond (p)

unit of force
10^{-3} kilopond (q.v.)

pound lb

UK and US unit, and FPS and ft-lbf-s base-unit of mass
Def: the pound equals to 0.453 592 37 kilogramme exactly (from 1959 in US, from 1963 in UK)
4.535 92 × 10^{-1} kg
4.625 35 × 10^{-2} metric technical unit of mass
3.108 10 × 10^{-2} slug
4.464 29 × 10^{-4} ton (UK or long)
5.0 × 10^{-4} short ton
1.6 × 10 oz
7.0 × 10^3 gr
Note: this is an avoirdupois unit

pound, troy (lb tr)

depr. unit of mass
3.732 41 × 10^{-1} kg

pound foot per second lb ft/s

FPS unit of momentum
1.382 55 × 10^{-1} kg m/s

pound foot per second squared lb ft/s^2

= poundal (q.v.)

pound foot squared lb ft^2

FPS unit of moment of inertia

$4.214\,01 \times 10^{-2}$ kg m^2

pound foot squared per second lb ft^2/s

FPS unit of moment of momentum

$4.214\,01 \times 10^{-2}$ kg m^2/s

pound inch squared lb in^2

unit of moment of inertia

$2.926\,40 \times 10^{-4}$ kg m^2

$6.944\,44 \times 10^{-3}$ lb ft^2

pound per acre lb/acre

unit of mass per unit area

$1.120\,85 \times 10^{-4}$ kg/m^2

$1.120\,85$ kg/ha

$2.295\,68 \times 10^{-2}$ lb/1000 ft^2

$3.305\,79 \times 10^{-3}$ oz/yd^2

$3.673\,09 \times 10^{-4}$ oz/ft^2

$2.857\,14 \times 10^{-1}$ ton/mile^2

pound per cubic foot lb/ft^3

FPS unit of (mass) density

$1.601\,85 \times 10$ kg/m^3

$5.787\,04 \times 10^{-4}$ lb/in^3

$1.605\,44 \times 10^{-1}$ lb/UKgal

$1.336\,81 \times 10^{-1}$ lb/USgal

pound per cubic inch lb/in^3

unit of (mass) density

$2.767\,99 \times 10^4$ kg/m^3

$\mathbf{1.728 \times 10^3}$ lb/ft^3

pound per foot lb/ft

unit of mass per unit length

1.488 16 kg/m

$8.333\,33 \times 10^{-2}$ lb/in

3.0 lb/yd

1.339 29 ton/1000 yd

pound per foot second lb/ft s

= poundal second per square foot (q.v.)

pound per gallon (UK) (lb/UKgal)

UK unit of (mass) density

$9.977\,64 \times 10$ kg/m^3

6.228 84 lb/ft^3

$8.326\,75 \times 10^{-1}$ lb/USgal

pound per gallon (US) (lb/USgal)

US unit of (mass) density

$1.198\,26 \times 10^2$ kg/m^3

7.480 52 lb/ft^3

1.200 95 lb/UKgal

pound per hour lb/h

unit of mass rate of flow

$1.259\,98 \times 10^{-4}$ kg/s

$4.535\,92 \times 10^{-1}$ kg/h

$2.777\,78 \times 10^{-4}$ lb/s

$4.464\,29 \times 10^{-4}$ ton/h

pound per inch lb/in

unit of mass per unit length

$1.785\,80 \times 10$ kg/m

$\mathbf{1.2 \times 10}$ lb/ft

$\mathbf{3.6 \times 10}$ lb/yd

$1.607\,14 \times 10$ ton/1000 yd

pound per second lb/s

unit of mass rate of flow

$4.535\,92 \times 10^{-1}$ kg/s

$1.632\,93 \times 10^3$ kg/h

$\mathbf{3.6 \times 10^3}$ lb/h

1.607 14 ton/h

pound per square foot lb/ft^2

unit of mass per unit area

4.882 43 kg/m^2

Note: this name and symbol are often used incorrectly for pound-force per square foot (q.v.)

pound per square inch lb/in^2

unit of mass per unit area

7.030 70 × 10^2 kg/m^2

Note: this name and symbol are often used incorrectly for pound-force per square inch (q.v.)

pound per thousand square feet lb/1000 ft^2

unit of mass per unit area

4.882 43 × 10^{-3} kg/m^2

4.882 43 × 10 kg/ha

4.356 × 10 lb/acre

1.44 × 10^{-1} oz/yd^2

1.6 × 10^{-2} oz/ft^2

1.244 57 × 10 ton/mile^2

pound per yard lb/yd

unit of mass per unit length

4.960 55 × 10^{-1} kg/m

2.777 78 × 10^{-2} lb/in

3.333 33 × 10^{-1} lb/ft

4.464 29 × 10^{-1} ton/1000 yd

poundal pdl

FPS unit of force

= lb ft/s^2

1.382 55 × 10^{-1} N

3.108 10 × 10^{-2} lbf

poundal foot pdl ft

FPS unit of moment of force

4.214 01 × 10^{-2} N m

3.108 01 × 10^{-2} lbf ft

poundal per square foot pdl/ft^2

FPS unit of pressure

1.488 16 N/m^2

1.517 50 $\times$ 10^{-5} kgf/cm^2 or kp/cm^2

3.108 10 $\times$ 10^{-2} lbf/ft^2

poundal second per square foot pdl s/ft^2

FPS unit of (dynamic) viscosity

1.488 16 N s/m^2

1.0 lb/ft s

pound-force (lbf)

ft-lbf-s base-unit of force

4.448 22 N

4.535 92 $\times$ 10^{-1} kgf

3.217 40 $\times$ 10 pdl

4.464 29 $\times$ 10^{-4} tonf

pound-force foot (lbf ft)

ft-lbf-s unit of moment of force

1.355 82 N m

1.382 55 $\times$ 10^{-1} kgf m or kp m

pound-force hour per square foot (lbf h/ft^2)

unit of (dynamic) viscosity

1.723 69 $\times$ 10^5 N s/m^2

pound-force inch (lbf in)

unit of moment of force

1.129 85 $\times$ 10^{-1} N m

2.681 17 pdl ft

8.333 33 $\times$ 10^{-2} lbf ft

pound-force per foot (lbf/ft)

ft-lbf-s unit of surface tension

1.459 39 $\times$ 10 N/m

pound-force per inch (lbf/in)

unit of surface tension

1.751 27 $\times$ 10^2 N/m

pound-force per square foot (lbf/ft^2)
ft-lbf-s unit of pressure
$4.788\,03 \times 10$ N/m^2
$4.882\,43 \times 10^{-4}$ kgf/cm^2 or kp/cm^2
$6.944\,44 \times 10^{-3}$ lbf/in^2

pound-force per square inch (lbf/in^2)
unit of pressure
$6.849\,76 \times 10^3$ N/m^2
$7.030\,7 \times 10^{-2}$ kgf/cm^2 or kp/cm^2
1.44 $\times$ **10^2** lbf/ft^2

pound-force second per square foot (lbf s/ft^2)
ft-lbf-s unit of (dynamic) viscosity
$4.788\,03 \times 10$ N s/m^2
1.0 slug/ft s

pound-weight (Lb)
Note: other name for pound-force (q.v.)

PS
G abbr. for *Pferdestärke* = metric horsepower (—)

p.s.f.; psf
'pounds' per square foot, correctly: pounds-force per square foot (lbf/ft^2)

p.s.i.; psi
'pounds' per square inch, correctly: pounds-force per square inch (lbf/in^2)

psia
'pounds' per square inch (absolute)—incorrect; instead of 'pressure of *x* psia' write 'absolute pressure of *x* lbf/in^2'

psig
pounds per square inch (gauge)—incorrect; instead of 'pressure of *x* psig' write 'gauge pressure of *x* lbf/in^2'

pt, pt. pint

pz pièze

Q

q quintal

qt UK quart

quart (UK) (UKqt)
UK unit of volume (capacity)
$1.136\,52 \times 10^{-3}$ m³

quart see: dry quart, liquid quart

quarter (—)
UK unit of mass
$1.270\,06 \times 10$ kg
2.8 × **10** lb

quintal (q)
quintal (q)
Dezitonne (dt)
unit of mass
$\mathbf{10^2}$ kg

R

R röntgen

°R degree Rankine; (degree Réaumur)

rad rad; radian

rad rad
rad
rad
unit of absorbed dose
$\mathbf{10^{-2}}$ J/kg = $\mathbf{10^2}$ erg/g
Note: the symbol for this unit is the same as the symbol of the unit radian

rad per second rad/s
unit of absorbed dose rate
$\mathbf{10^{-2}}$ W/kg = $\mathbf{10^2}$ erg/g s

radian rad

radian

Radiant

SI supplementary unit of plane angle

Def: the radian is the angle between two radii of a circle which cut off on the circumference an arc equal in length to the radius.

57° 17′ 44.8″

$5.729\ 58 \times 10$	°	(degree)	$(= 180/\pi)$
$3.437\ 75 \times 10^3$	′	(minute)	
$2.062\ 65 \times 10^5$	″	(second)	(206 264.806″)
$6.366\ 20 \times 10$	g	(grade)	$(= 200/\pi)$
$6.366\ 20 \times 10^{-1}$	∟	(right angle)	

Note: symbol rd is sometimes used in France

radian per minute rad/min

unit of angular velocity

$1.666\ 67 \times 10^{-2}$ rad/s

$9.549\ 30 \times 10^{-1}$ °/s

radian per second rad/s

SI unit of angular velocity

6.0 $\times$ **10** rad/min

$5.729\ 58 \times 10$ °/s

$6.366\ 20 \times 10$ g/s

Note: symbol rd/s is sometimes used in France

radian per second squared rad/s^2

SI unit of angular acceleration

$5.729\ 58 \times 10$ °/s^2

$6.366\ 20 \times 10$ g/s^2

rd F. obsol. symbol for radian

rd; Rd rutherford

reciprocal angström Å^{-1}

unit of wave number

10^{10} m^{-1} = **10^8** cm^{-1}

reciprocal centimetre cm^{-1}

CGS unit of wave number

$\mathbf{10^2\ m^{-1} = 10^{-8}\ Å^{-1}}$

Note: used in spectroscopy

reciprocal cubic metre m^{-3}

SI unit of number density of molecules (or particles)

reciprocal degree deg^{-1} ($degC^{-1}$; $degK^{-1}$)

unit of linear expansion coefficient

1 K^{-1} or $°C^{-1}$

reciprocal henry H^{-1}

SI unit of reluctance

$= m^{-2}\ kg^{-1}\ s^2\ A^2 = A/Wb = A/V\ s = 1/(\Omega\ s)$

$1.129\ 41 \times 10^{13}$ CGSe unit

$1.256\ 64 \times 10^{-8}$ Gb/Mx

reciprocal kelvin K^{-1}

SI unit of linear expansion coefficient

1.0 $°C^{-1}$ or deg^{-1}

$5.555\ 55 \times 10^{-1}$ $°F^{-1}$ or $degF^{-1}$

Note: also unit of cubic expansion coefficient and pressure coefficient

reciprocal metre m^{-1}

SI unit of wave number

1 dioptre = $\mathbf{10^{-2}}$ cm^{-1}

Note: also unit of attenuation coefficient, phase coefficient, propagation coefficient, linear absorption coefficient, Rydberg constant, macroscopic cross-section (cross-section density) and linear ionization of a particle

reciprocal minute min^{-1}

unit of angular frequency

$1.666\ 67 \times 10^{-2}\ s^{-1}$

reciprocal mole mol^{-1}

(SI) unit of molar volume

reciprocal ohm Ω^{-1}

SI unit of conductance

Note: the name siemens (symbol S) is adopted by ISO and IEC for this unit, and also used in this dictionary

reciprocal ohm metre $1/(\Omega\ m)$

= siemens per metre (q.v.)

reciprocal poise P^{-1}

CGS unit of fluidity

10 $m^2/N\ s$

Note: the name rhe was sometimes used for this unit

reciprocal second s^{-1}

SI unit of angular frequency

60 min^{-1}

Note: also unit of rotational frequency, pulsatance (circular frequency), damping coefficient, activity and decay constant

reciprocal second reciprocal cubic metre $s^{-1}\ m^{-3}$

SI unit of total neutron source density and slowing down density

reciprocal second reciprocal kilogramme $s^{-1}\ kg^{-1}$

SI unit of specific activity

reciprocal second reciprocal square metre $s^{-1}\ m^{-2}$

SI unit of particle flux density

= reciprocal square metre reciprocal second (q.v.)

reciprocal square metre reciprocal second $m^{-2}\ s^{-1}$

SI unit of current density of particles

= reciprocal second reciprocal square metre (q.v.)

rep (—)

obsol. unit of absorbed dose

93 erg/g

Note: short for *r*öntgen *e*quivalent *p*hysical

rev revolution

revolution (rev)

tour (tr)

Umdrehung (U)

unit of plane angle

6.283 19 rad

3.6 × 10^2 ° (degree)

4.0 × 10^2 ᵍ (grade)

revolution per minute (rev/min)

tour par minute (tr/min)

Umdrehung/Minute (U/min)

unit of rotational frequency

Note: symbol tr/mn is sometimes used in France

revolution per second (rev/s)

tour par seconde (tr/s)

Umdrehung/Sekunde (U/s)

unit of rotational frequency

rhe (—)

unit of fluidity

= reciprocal poise (q.v.)

right angle . . . . ∟

angle droit

Rechter; *rechter Winkel*

oSI unit of plane angle

1.570 80 rad (= $\pi/2$)

9.0 × **10** ° (degree)

5.4 × 10^3 ′ (minute)

3.24 × 10^5 ″ (second)

1.0 × 10^2 ᵍ (grade)

rod (—)

depr. unit of length

5.029 2 m = **5.5** yd

Note: also called pole or perch

röntgen R

röntgen

Röntgen

unit of ionization exposure

2.58 × 10^{-4} C/kg

röntgen equivalent physical

= rep (q.v.)

röntgen per second R/s

unit of ionization exposure rate

2.58 × 10^{-4} C/kg s

röntgen metre squared per curie hour R m^2/Ci h

unit of specific gamma ray constant

1.934 46 × 10^{-12} C m^2/kg

rood (—)

UK unit of area

1.011 71	× 10^3	m^2
1.21	**× 10^3**	yd^2
2.5	**× 10^{-1}**	acre

r.p.h. revolutions per hour (rev/h)

r.p.m. revolutions per minute (rev/min)

r.p.s. revolutions per second (rev/s)

rutherford (Rd)

obsolete unit of activity

10^6 s^{-1} = 2.702 70 × 10^{-5} Ci

Ry rydberg

rydberg (Ry)

2.179 72 × 10^{-18} J

1.360 54 × 10 eV

S

s second; stat

S siemens

sabin (—)
unit of equivalent absorption area
1 $ft^2 \approx 9.290\,30 \times 10^{-2}$ m^2

sb stilb

scruple (—)
UK and US unit of mass
$1.295\,98 \times 10^{-3}$ kg
2.0 × **10** gr
Note: this is an apothecaries' unit; can be used for drugs only

sec second (s)

secohm name sometimes used for ohm second (q.v.)

second s
seconde
Sekunde
SI base-unit of time
Def: the second is the unit of time defined under that name by the CGPM
$1.666\,67 \times 10^{-2}$ min
$2.777\,78 \times 10^{-4}$ h
$1.157\,41 \times 10^{-5}$ d
Note: also unit of mean life and half-life

second . . . ″
seconde
Sekunde; *Altsekunde*
oSI unit of plane angle
$4.848\,14 \times 10^{-6}$ rad
$2.777\,78 \times 10^{-4}$ ° (degree)
$1.666\,67 \times 10^{-2}$ ′ (minute)
$3.086\,42 \times 10^{-4}$ ^{g} (grade)
$3.086\,42 \times 10^{-6}$ ∟ (right angle)
Note: the second can be subdivided decimally

second see: centesimal second

second squared per kilogramme s^2/kg
= square metre per joule (q.v.)

Sek. G abbr for *Sekunde* = second (s)

sh cwt short hundredweight

short hundredweight (sh cwt)
US unit of mass
$4.535\,92 \times 10$ kg
1.0 $\times$ **10^2** lb
$8.928\,57 \times 10^{-1}$ hundredweight (UK)

short ton (sh tn)
US unit of mass
$9.071\,85 \times 10^2$ kg
2.0 $\times$ **10^3** lb
$8.928\,57 \times 10^{-1}$ ton (UK)
Note: in the US usually referred to as ton if there is no danger of confusion with long ton

sh tn short ton

Siegbahn unit see: X unit

siemens S
siemens
Siemens
(SI) unit of conductance, admittance, modulus of admittance, susceptance
$= m^{-2}\,kg^{-1}\,s^3\,A^2 = V/A = \Omega^{-1}$
$8.987\,55 \times 10^{11}$ sS or $s\Omega^{-1}$
$1.000\,00 \times 10^{-9}$ aS or $a\Omega^{-1}$
Note: cf. reciprocal ohm

siemens metre per square millimetre $S\,m/mm^2$
unit of conductivity
10^6 S/m

siemens per metre S/m

(SI) unit of conductivity

$= m^{-3}\,kg^{-1}\,s^3\,A^2$

$8.987\,55 \times 10^9$ sS/cm

$1.000\,00 \times 10^{-11}$ aS/cm

siemens square metre per mole S m^2/mol

(SI) unit of molar conductivity

sigma (σ)

depr. unit of time

10^{-6} s = 1 μs

sk skot

skot (sk)

obsol. unit of scotopic luminance

10^{-3} asb (apostilb)

slug (—)

ft-lbf-s unit of mass

$1.459\,39 \times 10$ kg

$3.217\,40 \times 10$ lb (= 9.806 65/0.3048)

1.488 16 metric technical unit of mass

Note: sometimes called gee pound

slug foot squared (slug ft^2)

ft-lbf-s unit of moment of inertia

1.355 82 kg m^2

slug per cubic foot (slug/ft^3)

ft-lbf-s unit of (mass) density

$5.153\,79 \times 10^2$ kg/m^3

sn sthène

sone (—)

sone

Sone

(SI) unit of loudness

sp spat

spat (sp)
unit of solid angle
1.256 64 × 10 sr (= 4π)
Note: 1 sp = the solid angle of the sphere

sq.; sq UK and US abbr. for square

square centimetre cm^2
mSI and CGS unit of area
1.0 × 10^{-4} m^2
1.55 000 × 10^{-1} in^2

square centimetre per dyne cm^2/dyn
CGS unit of compressibility
10 m^2/N

square centimetre per erg cm^2/erg
CGS unit of spectral cross section
10^3 m^2/J

square centimetre per kilogramme-force (cm^2/kgf)
unit of compressibility
1.019 72 × 10^{-5} m^2/N
7.030 70 × 10^3 in^2/lbf

square centimetre per steradian erg $cm^2/sr\ erg$
CGS unit of spectral differential cross section
10^3 $m^2/sr\ J$

square chain (—)
depr. unit of area
4.046 86 × 10^2 m^2
4.84 × 10^2 yd^2

square degree ($\square^\circ$)
depr. unit of solid angle
3.046 17 × 10^{-4} sr
1.234 57 × 10^{-4} $\square^g$ (square grade)

square foot ft^2
UK and US unit of area
9.290 30 × 10^{-2} m^2
1.44 × 10^2 in^2

square foot per hour ft^2/h
= foot squared per hour (q.v.)

square foot hour degree Fahrenheit per British thermal unit foot
(ft^2 h degF/Btu ft)
unit of thermal resistivity
$5.777\,89 \times 10^{-1}$ m K/W

square foot hour degree Fahrenheit per British thermal unit inch
(ft^2 h degF/Btu in)
unit of thermal resistivity
6.933 47 m K/W

square foot per poundal ft^2/pdl
FPS unit of compressibility
$6.719\,69 \times 10^{-1}$ m^2/N

square foot per pound-force (ft^2/lbf)
ft-lbf-s unit of compressibility
$2.088\,54 \times 10^{-2}$ m^2/N

square foot per second ft^2/s
= foot squared per second (q.v.)

square foot per ton-force ($ft^2/tonf$)
UK unit of compressibility
$9.323\,85 \times 10^{-6}$ m^2/N

square grade ($\square^g$)
depr. unit of solid angle
$2.467\,40 \times 10^{-4}$ sr
8.1 $\times$ **10**$^{-1}$ $\square^\circ$ (square degree)

square inch in^2
UK and US unit of area
6.451 6 $\times$ **10**$^{-4}$ m^2
6.451 6 cm^2
$7.716\,05 \times 10^{-4}$ yd^2
$6.944\,44 \times 10^{-3}$ ft^2
$1.273\,24 \times 10^{6}$ circular mil

5

square inch per pound-force (in^2/lbf)

unit of compressibility

$1.450\,38 \times 10^{-4}$ m^2/N

square inch per ton-force ($in^2/tonf$)

UK unit of compressibility

$6.474\,90 \times 10^{-8}$ m^2/N

square kilometre km^2

mSI unit of area

$\mathbf{10^6}$ m^2 = $\mathbf{10^4}$ a = $\mathbf{10^2}$ ha

$3.861\,02 \times 10^{-1}$ $mile^2$

$2.471\,05 \times 10^2$ acre

square metre m^2

SI unit of area

$\mathbf{10^{-4}}$ ha = $\mathbf{10^{-2}}$ a = $\mathbf{10^{28}}$ b (barn)

$3.861\,02 \times 10^{-7}$ $mile^2$

$2.471\,05 \times 10^{-4}$ acres

$1.195\,99$ yd^2

$1.076\,39 \times 10$ ft^2

$1.550\,00 \times 10^3$ in^2

Note: also unit of equivalent absorption area, nuclear quadrupole moment, cross section and atomic attenuation coefficient. Note that e.g., 1 square kilometre (1 km^2) is not 10^3 m^2 but $(10^3\ m)^2 = 10^6\ m^2$

square metre per joule m^2/J

SI unit of spectral cross section

$= kg^{-1}\ s^2 = s^2/kg$

$\mathbf{10^{-3}}$ cm^2/erg = $\mathbf{10^{21}}$ b/erg

square metre per kilogramme m^2/kg

SI unit of mass attenuation coefficient and mass energy transfer coefficient

Note: also unit of specific surface

square metre per kilogramme-force second ($m^2/kgf\ s$)

m-kgf-s unit of fluidity

$1.019\,72 \times 10^{-1}$ $m^2/N\ s$

square metre per mole m^2/mol
(SI) unit of molar absorption coefficient

square metre per newton m^2/N
SI unit of compressibility
1.0 × **10^{-1}** cm^2/dyn
9.806 65 × 10^4 cm^2/kgf or cm^2/kp
1.488 16 ft^2/pdl
4.788 03 × 10 ft^2/lbf
6.894 76 × 10^3 in^2/lbf

square metre per newton second $m^2/N\ s$
(SI) unit of fluidity
$10^{-1}\ P^{-1}$ = **9.806 65** $m^2/kgf\ s$

square metre per second m^2/s
= metre squared per second (q.v.)

square metre per steradian m^2/sr
SI unit of differential cross section
10^{28} b/sr

square metre per steradian joule $m^2/sr\ J$
SI unit of spectral differential cross section
10^{-3} cm^2/sr erg
10^{21} b/sr erg

square metre per volt second $m^2/V\ s$
SI unit of mobility
= $kg^{-1}\ s^2\ A = m^2/Wb$

square metre per weber m^2/Wb
= square metre per volt second (q.v.)

square micrometre μm^2
mSI unit of area
10^{-12} m
Note: also called *square micron*

square micron (μ^2)
micron carré
Quadratmikron
= square micrometre (q.v.)

square mile   mile^2

UK and US unit of area

$2.589\,99 \times 10^6$ m^2

2.589 99 km^2

6.4 $\times$ **10^2** acres

2.56 $\times$ **10^3** roods

3.097 6 $\times$ **10^6** yd^2

Note: the symbol mi^2 is used in the USA

square mile per ton   (mile^2/ton)

UK unit of specific surface

$2.549\,08 \times 10^3$ m^2/kg

$2.549\,08 \times 10^2$ ha/t

square millimetre   mm^2

mSI unit of area

1.0 $\times$ **10^{-6}** m^2

$1.550\,00 \times 10^{-3}$ in^2

$1.973\,53 \times 10^3$ circular mil

square minute   (□′)

depr. unit of solid angle

$2.777\,78 \times 10^{-4}$ □° (square degree)

square second   (□″)

depr. unit of solid angle

$2.777\,78 \times 10^{-4}$ □′ (square minute)

square yard   yd^2

UK and US unit of area

$8.361\,27 \times 10^{-1}$ m^2

9.0 ft^2

1.296 $\times$ **10^3** in^2

square yard per ton   (yd^2/ton)

UK unit of specific surface

$8.229\,22 \times 10^{-1}$ m^2/t

sr   steradian

st   stère

St stokes

standard atmosphere atm

atmosphère normale
physikalische Atmosphäre
unit of pressure
1.013 25 × 10⁵ N/m²

$1.013\,25 \times 10^5$	N/m²
1.013 25	bar
1.033 23	kgf/cm²
7.6×10^2	torr
$1.469\,59 \times 10$	lbf/in²

standard gravity see: p. 182

standard temperature and pressure (s.t.p.)

If not specified otherwise
standard temperature = 0°C
standard pressure = 1 atm

stat (s)

a prefix denoting a CGSe unit (used in US)

statampere (sA)

CGSe unit of electric current
$3.335\,64 \times 10^{-10}$ A

statampere centimetre squared (sA cm²)

CGSe unit of electromagnetic moment
$3.335\,64 \times 10^{-14}$ A m²

statampere per square centimetre (sA/cm²)

CGSe unit of current density
$3.335\,64 \times 10^{-6}$ A/m²

statcoulomb (sC)

CGSe unit of electric charge and electric flux
For electric charge:
$3.335\,64 \times 10^{-10}$ C
For electric flux:
$2.654\,42 \times 10^{-11}$ C

statcoulomb centimetre (sC cm)

CGSe unit of electric dipole moment

$3.335\ 64 \times 10^{-12}$ C m

statcoulomb per cubic centimetre (sC/cm^3)

CGSe unit of volume density of charge

$3.335\ 64 \times 10^{-4}$ C/m^3

statcoulomb per square centimetre (sC/cm^2)

CGSe unit of surface density of charge, electric polarization and displacement

For surface density and polarization:

$3.335\ 64 \times 10^{-6}$ C/m^2

For displacement:

$2.654\ 42 \times 10^{-7}$ C/m^2

statfarad (sF)

CGSe unit of capacitance

$1.112\ 65 \times 10^{-12}$ F

stathenry (sH)

CGSe unit of self inductance, mutual inductance and permeance

For inductances:

$8.987\ 55 \times 10^{11}$ H

For permeance:

$1.129\ 41 \times 10^{13}$ H

statohm (sΩ)

CGSe unit of resistance

$8.987\ 55 \times 10^{11}$ Ω

statohm centimetre (sΩ cm)

CGSe unit of resistivity

$8.987\ 55 \times 10^{9}$ Ω m

statsiemens (sS)

CGSe unit of conductance

$1.112\ 65 \times 10^{-12}$ S

statsiemens per centimetre (sS/cm)
CGSe unit of conductivity
$1.112\,65 \times 10^{-10}$ S/m

statvolt (sV)
CGSe unit of electric potential
$2.997\,92 \times 10^{2}$ V

statvolt per centimetre (sV/cm)
CGSe unit of electric field strength
$2.997\,92 \times 10^{4}$ V/m

steradian sr
stéradian
Steradiant
SI supplementary unit of solid angle
Def: the steradian is the solid angle which, having its vertex in the centre of a sphere, cuts off an area of the surface of the sphere equal to that of a square having sides of length equal to the radius of the sphere
$7.957\,75 \times 10^{-2}$ spat (= $1/4\pi$)
$3.282\,81 \times 10^{3}$ $\square^{\circ}$ (square degree)
$4.052\,85 \times 10^{3}$ $\square^{g}$ (square grade)

stère (st)
stère
Raummeter (rm); *Ster*
unit of volume (for timber only)
= 1 m^3

sthène (sn)
MTS unit of force
= t m/s^2
10^3 N

sthène per square metre (sn/m^2)
= pièze (q.v.)

stilb (sb)
unit of luminance
10^4 cd/m^2

stokes St

stokes

Stokes

CGS unit of kinematic viscosity

= cm^2/s

10^{-4} metre squared per second (q.v.)

stone (—)

UK unit of mass

6.350 29 kg

1.4 × **10** lb

survey foot (—)

US unit of length

Def: 1 US survey foot = 1200/3937 m

1.000 002 ft

Note: used for Coast and Geodetic surveys within the USA

T

t tonne

T tera; tesla

technical atmosphere at

atmosphère technique

technische Atmosphäre

unit of pressure

9.806 65 × **10^4** N/m^2

9.806 65 × **10^{-1}** bar

1.0 kgf/cm^2 or kp/cm^2

9.678 41 × 10^{-1} atm

1.422 33 × 10 lbf/in^2

telegraph nautical mile (—)

obsol. unit of length

1.855 32 × 10^3 m

6.087 × **10^3** ft

tera (T)

a prefix denoting $\times 10^{12}$. Examples: terahertz (THz), terajoule (TJ), teraohm (TΩ), terawatt (TW)

tesla T

tesla

Tesla

SI unit of magnetic flux density and magnetic polarization

$= kg\ s^{-2}A^{-1} = Wb/m^2 = Vs/m^2$

For magnetic flux density:

$3.335\ 64 \times 10^{-7}$ CGSe unit

$1.000\ 00 \times 10^{4}$ Gs

For magnetic polarization:

$2.654\ 42 \times 10^{-8}$ CGSe unit

$7.957\ 75 \times 10^{2}$ CGSm unit

th thermie

therm (—)

UK unit of heat

10^5 Btu

therm per gallon (UK) (therm/UKgal)

UK unit of calorific value (volume basis)

$2.320\ 80 \times 10^{10}$ J/m^3

thermie (th)

unit of heat (energy)

10^6 cal_{15}

thou colloquial name for milli-inch (q.v.)

ton (UK) (UKton)

UK unit of mass

$1.016\ 05 \times 10^{3}$ kg

1.016 05 t

2.24 $\times$ **10^3** lb

1.12 short ton

Note: also called long ton or gross ton to distinguish it from short ton (q.v.)

ton, metric name sometimes used for tonne (q.v.)

ton; metric see: tonne

ton, long other name for ton (UK)

ton, gross other name for ton (UK)

ton, net other name for short ton (q.v.)

ton mile (—)
UK unit of mass carried × distance (traffic factor)
1.635 17 t km

ton mile per gallon (UK) (ton mile/UKgal)
UK unit of mass carried × distance/volume (traffic factor)
$3.596\,87 \times 10^{-1}$ t km/dm^3 or t km/l

ton of refrigeration (—)
unit of heat flow rate (refrigerating capacity)
$3.516\,85 \times 10^3$ W or J/s
2.88×10^5 Btu/d
1.2×10^4 Btu/h
2.0×10^2 Btu/min

ton per cubic yard (ton/yd^3)
UK unit of (mass) density
$1.328\,94 \times 10^3$ kg/m^3
$8.296\,30 \times 10$ lb/ft^3

ton per hour (ton/h)
UK unit of mass rate of flow
$2.822\,35 \times 10^{-1}$ kg/s
$1.016\,05 \times 10^3$ kg/h
$6.222\,22 \times 10^{-1}$ lb/s
2.24×10^3 lb/h

ton per mile (ton/mile)
UK unit of mass per unit length
$6.313\,42 \times 10^{-1}$ kg/m
$4.242\,42 \times 10^{-1}$ lb/ft
$5.681\,82 \times 10^{-1}$ ton/1000 yd

ton per square mile (ton/mile^2)

UK unit of mass per unit area

$3.922\,98 \times 10^{-4}$ kg/m^2

3.922 98 kg/ha

3.5 lb/acre

$8.034\,89 \times 10^{-2}$ lb/1000 ft^2

$1.157\,02 \times 10^{-2}$ oz/yd^2

$1.285\,58 \times 10^{-3}$ oz/ft^2

ton per thousand yards (ton/1000 yd)

UK unit of mass per unit length

1.111 16 kg/m

$6.222\,22 \times 10^{-2}$ lb/in

$7.466\,67 \times 10^{-1}$ lb/ft

2.24 lb/yd

tonf ton-force

ton-force (tonf)

UK unit of force

$9.964\,02 \times 10^{3}$ N

$7.206\,99 \times 10^{4}$ pdl

$\mathbf{2.24} \times \mathbf{10}^{3}$ lbf

ton-force foot (tonf ft)

UK unit of moment of force

$3.037\,03 \times 10^{3}$ N m

$3.096\,91 \times 10^{2}$ kgf m or kp m

$\mathbf{2.24} \times \mathbf{10}^{3}$ lbf ft

ton-force per foot (tonf/ft)

UK unit of force per unit length

$3.269\,03 \times 10^{4}$ N/m

ton-force per square foot (tonf/ft^2)

UK unit of pressure

$1.072\,52 \times 10^{5}$ N/m^2

1.093 66 kgf/cm^2 or kp/cm^2

$\mathbf{2.24} \times \mathbf{10}^{3}$ lbf/ft^2

ton-force per square inch (tonf/in^2)

UK unit of pressure

$1.544\ 43 \times 10^7$ N/m^2

$1.574\ 88 \times 10^2$ kgf/cm^2 or kp/cm^2

2.24 × **10**^{3} lbf/in^2

tonne t

tonne

Tonne

oSI unit and MTS base-unit of mass

10^3 kg = **1** Mg (megagramme)

Note: also called metric ton

tonne kilometre t km

unit of mass carried × distance (traffic factor)

$6.115\ 58 \times 10^{-1}$ ton mile

tonne metre per second squared t m/s^2

= sthène (q.v.)

tonne per cubic metre t/m^3

oSI and MTS unit of (mass) density

10^3 kg/m^3 = 1 Mg/m^3 or kg/dm^3

tonne per hectare t/ha

unit of mass per unit area

10 kilogramme per square metre (q.v.)

torr (torr)

torr

Torr

unit of pressure

$1.333\ 22 \times 10^2$ N/m^2

$1.333\ 22 \times 10^{-3}$ bar

$1.359\ 51 \times 10$ kgf/m^2 or kp/m^2

1.000 00 mmHg (conventional)

$1.315\ 79 \times 10^{-3}$ atm

$1.933\ 68 \times 10^{-2}$ lbf/in^2

torr litre per second (torr l/s)
unit of leak rate used in vacuum technology
$1.333\,22 \times 10^{-1}$ N m/s
1.0 $\times$ **10^{-3}** lusec

tr F abbr. for *tour* = revolution (rev)

trol troland

troland (trol)
unit of retinal illumination
1 trol = retinal illumination produced by a surface having luminance $L = 1$ cd/m^2 when the area of the retina $A = 1$ mm^2. In Germany it is considered to be a unit of retinal luminous intensity (*Pupillen-Lichtstärke* —I_p), cf. DIN 5031, Blatt 6

tropical year a
unit of time
365.242 198 78 d at 1900.0 decreasing at the rate of 0.000 006 14 d per century
Note: this is the year on which the calendar is based. It is the time between two consecutive passages (in the same direction) of the sun through the earth's equatorial plane

troy units
Obsolescent units of mass used rarely in US and UK and including the following:

1 troy pound (lb tr)	= **12** troy ounces
1 troy ounce (oz tr)	= **20** pennyweights
1 pennyweight (dwt)	= **24** grains
1 grain (gr) (no symb. in US)	= **1/480** troy ounce

U

u atomic mass unit (unified)

U G abbr. for *Umdrehung* = revolution (rev)

ua; UA microampere (μA)

u.e.m. F abbr. for *unité électromagnétique*
= electromagnetic unit (e.m.u.)

u.e.s. F abbr. for *unité électrostatique*
= electrostatic unit (e.s.u.)

UK
indicates a unit used in the United Kingdom, having the same or similar name as but a different value than a US unit, e.g. UK fl dr, UK fl oz, UKgal, UKmin, UKpt, UKqt

US
indicates a unit used in the United States of America, having the same name as but a different value than a UK unit, e.g. US fl dr, US fl oz, USgal, USmin

V

var var
var
Var
(SI) unit of reactive power
1 W

V volt

VA volt ampere

volt V
volt
Volt
SI unit of electric potential, potential difference and electromotive force
$= m^2 kg\, s^{-3} A^{-1} = W A^{-1} = A\Omega$
$3.335\,64 \times 10^{-3}$ sV
$1.000\,00 \times 10^{8}$ aV

volt ampere VA
SI unit of apparent power
1 W

volt per ampere V/A
= ohm (q.v.)

volt per metre V/m
SI unit of electric field strength
= m kg s^{-3} A^{-1}
3.335 64 × 10^{-5} sV/cm
1.000 00 × 10^{6} aV/cm

volt per mil (V/mil)
unit of electric field strength
3.937 01 × 10^{4} V/m

volt second V s
= weber (q.v.)

volt second per ampere V s/A
= henry (q.v.)

volt second per ampere metre V s/A m
= henry per metre (q.v.)

volt second per square metre V s/m^2
= tesla (q.v.)

W

W watt

watt W
watt
Watt
SI unit of power
= m^2 kg/s^3 = J/s = VA
1.019 72 × 10^{-1} kgf m/s
1.359 62 × 10^{-3} metric horsepower
7.375 62 × 10^{-1} ft lbf/s
1.341 02 × 10^{-3} hp (British horsepower)
Note: also unit of heat flow rate, radiant flux and sound energy flux

watt per ampere squared W/A^2
= ohm (q.v.)

watt per centimetre degree Celsius W/cm degC
$= 10^2$ watt per metre kelvin (q.v.)

watt per cubic foot W/ft^3
unit of heat release
3.531 47 $\times$ 10 W/m^3

watt per cubic metre W/m^3
(SI) unit of heat release (rate of heat liberation per unit volume)
$= m^{-1} kg\ s^{-3}$
8.598 45 $\times$ 10^{-1} kcal/m^3 h
1.0 $\times$ **10** erg/cm^3 s
9.662 11 $\times$ 10^{-2} Btu/ft^3 h

watt per foot degree Celsius (W/ft degC)
unit of thermal conductivity
3.280 84 W/m K

watt per kilogramme W/kg
SI unit of absorbed dose rate and kerma rate
$= m^2 s^{-3}$ = J/kg s
10^4 erg/g s = **10^2** rad/s

watt per metre degree Celsius or Kelvin
W/m degC or W/m degK
= watt per metre kelvin (q.v.)

watt per metre kelvin W/m K
SI unit of thermal conductivity
$= m\ kg\ s^{-3} K^{-1}$ = W/m degC
= J m/m^2 s K = J/m s K = J/m s degC
1.0 $\times$ **10^5** erg/cm s degC
2.388 46 $\times$ 10^{-3} cal/cm s degC
8.598 45 $\times$ 10^{-1} kcal/m h degC
1.019 72 $\times$ 10^{-1} kgf/s degC or kp/s degC
5.777 89 $\times$ 10^{-1} Btu/ft h degF
6.933 47 Btu in/ft^2 h degF

watt per square centimetre W/cm^2
$= 10^4$ watt per square metre (q.v.)

watt per square foot W/ft^2

unit of heat flow rate
1.076 39 × 10 W/m^2
9.255 29 kcal/m^2 h
3.412 14 Btu/ft^2 h

watt per square inch W/in^2

unit of heat flow rate
1.550 00 × 10^3 W/m^2
1.332 76 × 10^3 kcal/m^2 h
4.913 48 × 10^2 Btu/ft^2 h

watt per square metre W/m^2

SI unit of heat flow rate
= kg/s^3 = J/m^2 s
9.290 30 × 10^{-2} W/ft^2
6.451 6 × 10^{-4} W/in^2
1.0 × 10^3 erg/cm^2 s
2.388 46 × 10^{-5} cal/cm^2 s
8.598 45 × 10^{-1} kcal/m^2 h
3.169 98 × 10^{-1} Btu/ft^2 h
Note: also unit of radiant flux density, radiant exitance, irradiance, sound intensity and energy flux density

watt per square metre degree Celsius (W/m^2 degC)

= watt per square metre kelvin (q.v.)

watt per square metre kelvin W/m^2 K

SI unit of coefficient of heat transfer
= kg s^{-3}K^{-1} = J/m^2 s K = W/m^2 degC
2.388 46 × 10^{-5} cal/cm^2 s degC
8.598 45 × 10^{-1} kcal/m^2 h degC
1.0 × 10^3 erg/cm^2 s degC
1.019 72 × 10^{-1} kgf/m s degC or kp/m s degC
1.761 10 × 10^{-1} Btu/ft^2 h degF

watt per square metre kelvin to the fourth power W/m^2 K^4

SI unit of Stefan-Boltzmann constant
= m^{-2} kg s^{-3} K^{-4}
10^3 erg/cm^2 s K^4

watt per steradian W/sr
SI unit of radiant intensity
$= m^2\ kg\ s^{-3}\ sr^{-1}$
$\mathbf{10^7}$ erg/s sr

watt per steradian square metre $W/sr\ m^2$
SI unit of radiance
$= kg\ s^{-3}\ sr^{-1}$
$\mathbf{10^3}$ erg/s sr cm^2

watt second Ws
= joule (q.v.)

watt square metre $W\ m^2$
SI unit of first radiation constant
$= m^4\ kg\ s^{-3}$
$\mathbf{10^{11}}$ $W\ m^2$

Wb weber

weber Wb
weber
Weber
SI unit of magnetic flux (flux of magnetic induction)
$= m^2\ kg/s^2\ A = Vs$
$3.335\ 64 \times 10^{-3}$ CGSe unit of magnetic flux
$1.000\ 00 \times 10^{8}$ Mx

weber metre Wb m
SI unit of magnetic dipole moment
$= m^3\ kg\ s^{-2}\ A^{-1} = N\ m^2/A$

weber per ampere Wb/A
= henry (q.v.)

weber per ampere metre Wb/A m
= henry per metre (q.v.)

weber per metre Wb/m
SI unit of magnetic vector potential
$= m\ kg\ s^{-2}\ A^{-1} = V\ s/m$

weber per square metre Wb/m^2
= tesla (q.v.)

week (—)
semaine
Woche
unit of time
6.048 $\times$ **10^5** s
1.68 $\times$ **10^2** h

X

X.U. X-unit

X unit (X.U.)
depr. unit of wave length
1.002 02 $\times$ 10^{-13} m
Note: also called Siegbahn unit

Y

yard yd
UK and US unit of length
Def: the yard equals to 0.9144 metre exactly (from 1959 in US, from 1963 in UK)
3 ft = **36** in

yard per pound yd/lb
UK and US unit of length per unit mass
2.015 91 m/kg

yd yard

year a
année
Jahr
unit of time
Note: see Julian year and tropical year

yr year (a)

γ gamma (q.v.)

μ micro; micron (q.v.)

μb microbar

μbar microbar

μHg

conventional micron of mercury = conventional micrometre of mercury (μmHg)

μin microinch

μm micrometre

μmHg conventional micrometre of mercury

μs microsecond

σ sigma (q.v.)

Ω ohm

. . . .$^{\circ}$ degree

. . . .$'$ minute; foot (ft); centigrade (. . . .^{cg})

. . . .$''$ second; inch (in); one hundredth of a centigrade (. . . .^{cc})

$\square^{\circ}$ square degree

$\square^{g}$ square grade

$^0/_0$ per cent; per hundred

$^0/_{00}$ per thousand

PART II

A Dictionary of Quantities and Constants

A

ABSOLUTE TEMPERATURE

obsol. name for thermodynamic temperature (q.v.)

ABSORBED DOSE D Dim: L^2T^{-2}

dose absorbée

Energiedosis

SI unit: joule per kilogramme J/kg

other unit: rad rad

ABSORBED DOSE RATE $\dot{D}$ Dim: L^2T^{-3}

débit de dose absorbée

Energiedosisleistung; (Energiedosisrate)

SI unit: watt per kilogramme W/kg

other unit: rad per second rad/s

ACCELERATION see: angular acceleration, linear acceleration, standard gravity

ACCELERATION (LINEAR ACCELERATION) a Dim: LT^{-2}

accélération

Beschleunigung

SI unit: metre per second squared m/s^2

ACCELERATION DUE TO GRAVITY see: acceleration of free fall

ACCELERATION OF FREE FALL g Dim: LT^{-2}

accélération de la pesanteur; accélération due à la pesanteur

Fallbeschleunigung

SI unit: metre per second squared m/s^2

Note: also called acceleration due to gravity

ACOUSTIC ABSORPTION COEFFICIENT α, (α_a) Dim: 1

facteur d'absorption acoustique

Schallabsorptionsgrad

ACOUSTIC IMPEDANCE Z_a, (Z) Dim: $L^{-4}MT^{-1}$

impédance acoustique

akustische Impedanz; Flussimpedanz

(SI) unit: newton second per metre to the fifth $N\ s/m^5$

ACOUSTIC POWER see: sound energy flux

ACTION (*H*) Dim: L^2MT^{-1}

action

Wirkung

(SI) unit: joule second J s

ACTIVE POWER *P* Dim: L^2MT^{-3}

puissance active

Wirkleistung

SI unit: watt W

ACTIVITY *A* Dim: T^{-1}

activité

Aktivität

SI unit: reciprocal second s^{-1}

other unit: curie Ci

ADMITTANCE *Y* Dim: $L^{-2}M^{-1}T^3I^2$

admittance

Scheinleitwert

SI unit: reciprocal ohm Ω^{-1} (also called mho)

(SI) unit: siemens S

AFFINITY *A* Dim: L^2MT^{-2}

affinité

Affinität

(SI) unit: joule per mole J/mol

AMOUNT OF SUBSTANCE n, (ν) Dim: 1

quantité de matière

Stoffmenge

(SI) unit: mole mol

Note: this quantity is treated in practice as a base-quantity

AMOUNT OF SUBSTANCE CONCENTRATION see: molarity

ANGLE see: solid angle

ANGLE (PLANE ANGLE) $\alpha, \beta, \gamma, \vartheta, \theta, \varphi$, etc. Dim: 1

angle; (angle plan)

Winkel; (ebener Winkel)

SI unit: radian rad

other units: degree, minute, second; grade

ANGULAR ACCELERATION α Dim: T^{-2}

accélération angulaire

Winkelbeschleunigung

SI unit: radian per second squared rad/s^2

ANGULAR FREQUENCY ω Dim: T^{-1}

pulsation

Kreisfrequenz

SI unit: reciprocal second s^{-1}

ANGULAR MOMENTUM $b, p_\theta, p_\vartheta; \boldsymbol{L}$ Dim: L^2MT^{-1}

moment cinétique; moment de quantité de mouvement

Drehimpuls

SI unit: kilogramme metre squared per second kg m^2/s

Note: also called moment of momentum

ANGULAR VELOCITY ω Dim: T^{-1}

vitesse angulaire

Winkelgeschwindigkeit

SI unit: radian per second rad/s

Note: cf. rotational frequency

APPARENT POWER $S, (P_s)$ Dim: L^2MT^{-3}

puissance apparente

Scheinleistung

SI unit: volt ampere V A

AREA $A, (S)$ Dim: L^2

aire, superficie

Fläche

SI unit: square metre m^2

other units: are a, hectare ha

ATOMIC ATTENUATION COEFFICIENT μ_a, μ_{at} Dim: L^2

coefficient atomique d'atténuation

atomarer Schwächungskoeffizient

SI unit: square metre m^2

ATOMIC MASS CONSTANT (UNIFIED) m_u Dim: M

constante (unifiée) de masse atomique

atomare Massenkonstante (vereinheitlichte)

$1.660\,44 \times 10^{-27}$ kg

1.0 u

ATOMIC NUMBER Z Dim: 1

nombre atomique; nombre de charge

Ordnungszahl; Ladungszahl

Note: also called proton number

ATOMIC STOPPING POWER S_a Dim: L^4MT^{-2}

pouvoir atomique d'arrêt

atomares Bremsvermögen

SI unit: joule square metre $J\ m^2$

other unit: electronvolt square centimetre $eV\ cm^2$

ATOMIC WEIGHT see: relative atomic mass

ATTENUATION COEFFICIENT α Dim: L^{-1}

constante d'affaiblissement

Dämpfungskoeffizient

SI unit: reciprocal metre m^{-1}

Note: neper per metre Np/m is sometimes used as a unit of this quantity

AVERAGE LOGARITHMIC ENERGY DEGREMENT ξ Dim: 1

décrément logarithmique moyen de l'énergie

mittleres logarithmisches Energiedekrement

AVOGADRO CONSTANT L, N_A Dim: 1

constante d'Avogadro

Avogadro-Konstante

$6.022\,52 \times 10^{23}\ mol^{-1}$

B

BENDING MOMENT M Dim: L^2MT^{-2}

moment de flexion

Biegemoment

SI unit: newton metre $N\ m$

BOHR MAGNETON μ_B Dim: L^2I

magnéton de Bohr

Bohrsches Magneton

9.2732×10^{-24} J/T or $A\ m^2$

BOHR RADIUS a_0 Dim: L

rayon de Bohr

Bohr-Radius; Radius der ersten Bohrschen Kreisbahn

$5.291\,67 \times 10^{-11}$ m

BOLTZMANN CONSTANT k Dim: $L^2MT^{-2}\Theta^{-1}$

constante de Boltzmann

Boltzmann-Konstante

$1.380\,54 \times 10^{-23}$ J/K

BREADTH b Dim: L

largeur

Breite

SI unit: metre m

BRIGHTNESS see: luminance

BULK COMPRESSIBILITY see: compressibility

BULK MODULUS K Dim: $L^{-1}MT^{-2}$

module de compressibilité volumique (sous pression hydrostatique); module de compression

Kompressionsmodul

SI unit: newton per square metre N/m^2

Note: also called modulus of compression

BULK STRAIN see: volume strain

C

CAPACITANCE C Dim: $L^{-2}M^{-1}T^4I^2$

capacité

elektrische Kapazität

SI unit: farad F

CELSIUS TEMPERATURE t, θ, ϑ Dim: Θ

température Celsius

Celsius-Temperatur

SI unit: degree Celsius °C

Note: cf. thermodynamic temperature and temperature interval

CHARGE OF POSITRON see: elementary charge

CHEMICAL POTENTIAL OF COMPONENT B μ_B Dim: L^2MT^{-2}
potentiel chimique du constituant B
chemisches Potential der Komponente B
(SI) unit: joule per mole J/mol

CIRCULAR WAVE NUMBER k Dim: L^{-1}
nombre d'onde circulaire
Kreiswellenzahl
SI unit: reciprocal metre m^{-1}

COEFFICIENT OF FRICTION μ, (f) Dim: 1
coefficient de frottement; facteur de frottement
Reibungszahl

COEFFICIENT OF HEAT TRANSFER h, K, U, α
Dim: $MT^{-3}\Theta^{-1}$
coefficient de transmission thermique
Wärmeübergangskoeffizient; Wärmeübergangszahl
SI unit: watt per square metre kelvin W/m^2 K
other unit: watt per square metre degree Celsius W/m^2 °C
Note: also called thermal conductance or heat transfer coefficient

COEFFICIENTS
see: acoustic absorption coefficient, atomic attenuation coefficient, attenuation coefficient, cubic expansion coefficient, damping coefficient, diffusion coefficient, dissipation coefficient, linear attenuation coefficient, linear expansion coefficient, mass attenuation coefficient, mass energy transfer coefficient, pressure coefficient, propagation coefficient, recombination coefficient, reflection coefficient, transmission coefficient

COMPRESSIBILITY κ Dim: $LM^{-1}T^2$
coefficient de compressibilité; (coefficient de compressibilité volumique sous pression hydrostatique)
Kompressibilität
SI unit: square metre per newton m^2/N
Note: also called bulk compressibility

COMPTON WAVELENGTH λ_C Dim: L

longueur d'onde de Compton
Compton-Wellenlänge
SI unit: metre m

CONCENTRATION OF MOLECULES OF COMPONENT B C_B, n_B
Dim: L^{-3}

concentration de molécules du constituant B
Molekülkonzentration der Komponente B
SI unit: reciprocal cubic metre m^{-3}

CONDUCTANCE G Dim: $L^{-2}M^{-1}T^3I^2$

conductance
elektrischer Leitwert
SI unit: reciprocal ohm Ω^{-1} (also called mho)
(SI) unit: siemens S
Note: reciprocal of conductance is called resistance

CONDUCTIVITY γ, σ Dim: $L^{-3}M^{-1}T^3I^2$

conductivité
elektrische Leitfähigkeit
SI unit: reciprocal ohm metre $1/\Omega$ m
(SI) unit: siemens per metre S/m
Note: reciprocal of conductivity is called resistivity

CONSTANTS

see: atomic mass constant (unified), Avogadro constant, Bohr magneton, Bohr radius, Boltzmann constant, electron radius, elementary charge, Faraday constant, fine-structure constant, first radiation constant, molar gas constant, nuclear magneton, permeability of vacuum, Planck constant, rest mass of electron, neutron and proton, Rydberg constant, second radiation constant, standard gravity, Stefan-Boltzmann constant, velocity of light *in vacuo*. Also: decay constant, reactor time constant, specific gamma ray constant

CROSS SECTION σ Dim: L^2

section efficace
Wirkungsquerschnitt
SI unit: square metre m^2
other unit: barn b

CUBIC EXPANSION COEFFICIENT α, β, γ Dim: Θ^{-1}

coefficient de dilatation volumique

Volumen-Ausdehnungskoeffizient

SI unit: reciprocal kelvin K^{-1}

other unit: reciprocal degree Celsius $°C^{-1}$

Note: also called volume expansion coefficient

CURRENT DENSITY OF PARTICLES $J, (S)$ Dim: $L^{-2}T^{-2}$

densité de courant de particules

Teilchenstromdichte

SI unit: reciprocal square metre reciprocal second $m^{-2} s^{-1}$

D

DAMPING COEFFICIENT δ Dim: T^{-1}

coefficient d'amortissement

Abklingkonstante

SI unit: reciprocal second s^{-1}

Note: neper per second Np/s is sometimes used as a unit of this quantity

DECAY CONSTANT λ Dim: T^{-1}

constante de désintégration

Zerfallskonstante

SI unit: reciprocal second s^{-1}

Note: also called disintegration constant

DENSITY (MASS DENSITY) ρ Dim: $L^{-3}M$

masse volumique

Dichte; (*Massendichte*)

SI unit: kilogramme per cubic metre kg/m^3

other units: t/m^3, kg/l, g/ml, g/l

Note: reciprocal of (mass) density is called specific volume

DENSITY OF HEAT FLOW RATE $q, (\varphi)$ Dim: MT^{-3}

densité de flux thermique

Wärmestromdichte

SI unit: watt per square metre W/m^2

Note: also called intensity of heat flow rate

DIAMETER d Dim: L

diamètre

Durchmesser

SI unit: metre m

DIFFERENTIAL CROSS SECTION σ_Ω Dim: L^2

section efficace différentielle

differentieller Wirkungsquerschnitt

SI unit: square metre per steradian m^2/sr

other unit: barn per steradian b/sr

DIFFUSION AREA L^2 Dim: L^2

aire de diffusion

Diffusionsfläche

SI unit: square metre m^2

DIFFUSION COEFFICIENT D Dim: L^2T^{-1}

coefficient de diffusion

Diffusionskoeffizient

SI unit: square metre per second m^2/s

DIFFUSION COEFFICIENT FOR NEUTRON FLUX DENSITY D_ϕ, (D) Dim: L

coefficient de diffusion pour la densité de flux de neutrons

Diffusionskoeffizient für Neutronenflussdichte

SI unit: metre m

DIFFUSION LENGTH L Dim: L

longueur de diffusion

Diffusionslänge

SI unit: metre m

DIRECTIONAL SPECTRAL EMISSIVITY $\varepsilon(\lambda, \theta, \varphi)$ Dim: 1

émissivité spectrale directionelle

gerichteter spektraler Emissionsgrad

DISINTEGRATION ENERGY Q Dim: L^2MT^{-2}

énergie de désintégration

Zerfallsenergie

SI unit: joule J

DISSIPATION COEFFICIENT δ Dim: 1

facteur de dissipation

Schalldissipationsgrad

DYNAMIC MOMENT OF INERTIA see: moment of inertia

DYNAMIC VISCOSITY see: viscosity

E

EFFICIENCY η Dim: 1
rendement
Wirkungsgrad

ELECTRIC CHARGE Q Dim: TI
charge électrique; quantité d'électricité
elektrische Ladung; Elektrizitätsmenge
SI unit: coulomb C
Note: also called quantity of electricity

ELECTRIC CURRENT I Dim: I
courant électrique; (intensité de courant électrique)
elektrische Stromstärke
SI unit: ampere A
Note: also called intensity of electric current

ELECTRIC CURRENT DENSITY $J, (S)$ Dim: $L^{-2}I$
densité de courant électrique
elektrische Stromdichte
SI unit: ampere per square metre A/m^2

ELECTRIC DIPOLE MOMENT $p, (p_e); \boldsymbol{p}$ Dim: LTI
moment de dipôle électrique
Dipolmoment
SI unit: coulomb metre C m

ELECTRIC DISPLACEMENT $D; \boldsymbol{D}$ Dim: $L^{-2}TI$
déplacement électrique
elektrische Flussdichte; (elektrische Verschiebung)
SI unit: coulomb per square metre C/m^2

ELECTRIC FIELD STRENGTH $E, (K); \boldsymbol{E}$ Dim: $LMT^{-3}I^{-1}$
champ électrique
elektrische Feldstärke
SI unit: volt per metre V/m

ELECTRIC FLUX Ψ Dim: TI

flux électrique; flux de déplacement
elektrischer Fluss
SI unit: coulomb C
Note: also called flux of displacement

ELECTRIC POLARIZATION P; $\boldsymbol{P}$ Dim: $L^{-2}TI$

polarisation électrique
elektrische Polarisation
SI unit: coulomb per square metre C/m^2

ELECTRIC POTENTIAL V, φ Dim: $L^2MT^{-3}I^{-1}$

potentiel électrique
elektrisches Potential
SI unit: volt V

ELECTRIC SUSCEPTIBILITY χ_e Dim: 1

susceptibilité électrique
elektrische Suszeptibilität

ELECTROMAGNETIC ENERGY DENSITY w Dim: $L^{-1}MT^{-2}$

densité d'énergie électromagnétique
Energiedichte des elektromagnetischen Feldes
SI unit: joule per cubic metre J/m^3

ELECTROMAGNETIC MOMENT m; $\boldsymbol{m}$ Dim: L^2I

moment électromagnétique
das Ampèrsche magnetische Moment
SI unit: ampere metre squared $A\,m^2$

ELECTROMOTIVE FORCE E Dim: $L^2MT^{-3}I^{-1}$

force électromotrice
elektromotorische Kraft
SI unit: volt V

ELECTRON RADIUS r_e Dim: L

rayon d'électron
Elektronenradius
$2.817\,77 \times 10^{-15}$ m

ELEMENTARY CHARGE e Dim: TI

charge élémentaire
Elementarladung
$1.602\,10 \times 10^{-19}$ C

EMISSIVITY ε Dim: 1

émissivité

Emissionsgrad

ENERGY E, W Dim: L^2MT^{-2}

énergie

Energie

SI unit: joule J

other units: kilowatt hour kWh; electronvolt eV

ENERGY FLUX DENSITY ψ Dim: MT^{-3}

densité de flux d'énergie

Energieflussdichte

SI unit: watt per square metre W/m^2

ENERGY IMPARTED (TO MATTER) E_D Dim: L^2MT^{-2}

énergie communiquée (à la matière)

(auf Materie) übertragene Energie

SI unit: joule J

other unit: gramme rad g rad

Note: also called integral absorbed dose

ENTHALPY $H, (I)$ Dim: L^2MT^{-2}

enthalpie

Enthalpie

SI unit: joule J

ENTROPY S Dim: $L^2MT^{-2}\Theta^{-1}$

entropie

Entropie

SI unit: joule per kelvin J/K

EQUIVALENT ABSORPTION AREA A Dim: L^2

aire d'absorption équivalente

äquivalente Absorptionsfläche

SI unit: square metre m^2

EXPOSURE X Dim: $M^{-1}TI$

exposition; exposition d'ionisation

Ionendosis

SI unit: coulomb per kilogramme C/kg

other unit: röntgen R

Note: also called ionization exposure

EXPOSURE RATE $\dot{X}$ Dim: $M^{-1}I$

débit d'exposition; débit d'exposition d'ionisation
Ionendosisleistung; (Ionendosisrate)
SI unit: coulomb per kilogramme second C/(kg s)
other unit: röntgen per second R/s
Note: also called ionization exposure rate

F

FARADAY CONSTANT F Dim: TI

constante de Faraday
Faraday-Konstante
$9.648\,70 \times 10^4$ C/mol

FAST FISSION FACTOR ε Dim: 1

facteur de fission rapide
Schnellspaltfaktor

FINE STRUCTURE CONSTANT α Dim: 1

constante de structure fine
Feinstrukturkonstante
$7.297\,20 \times 10^{-3}$
Note: $\alpha^{-1} \approx 1.370\,39 \times 10^2$
$\alpha^2 \approx 5.324\,92 \times 10^{-5}$

FIRST RADIATION CONSTANT c_1 Dim: L^4MT^{-3}

première constante de rayonnement
die erste Strahlungskonstante
3.7415×10^{-16} W m^2

FLUIDITY (φ) Dim: $LM^{-1}T$

fluidité
Fluidität; Beweglichkeit
SI unit: metre squared per newton second m^2/N s
Note: reciprocal of fluidity is called (dynamic) viscosity

FLUX OF DISPLACEMENT see: electric flux

FLUX OF MAGNETIC INDUCTION see: magnetic flux

FOCUS LENGTH — Dim: L

distance focale
Brennweite
SI unit: metre m
Note: also called focal length

FORCE *F*; ***F*** — Dim: LMT^{-2}

force
Kraft
SI unit: newton N

FREE ENERGY *F* — Dim: L^2MT^{-2}

énergie libre
freie Energie
SI unit: joule J
Note: also called Helmholtz function

FREE ENTHALPY see: Gibbs function

FREQUENCY see: rotational frequency

FREQUENCY *f*, *ν* — Dim: T^{-1}

fréquence
Frequenz
SI unit: hertz Hz

FREQUENCY INTERVAL (*i*) — Dim: 1

intervalle de fréquence
Frequenzintervall
(SI) unit: octave (—)

G

GEOMETRICAL MOMENT OF INERTIA see: second moment of area

GIBBS FUNCTION *G* — Dim: L^2MT^{-2}

enthalpie libre; fonction de Gibbs
freie Enthalpie; Gibbs-Funktion
SI unit: joule J

GRAVITATIONAL CONSTANT *G* — Dim: $L^3M^{-1}T^{-2}$

constante de gravitation
Gravitationskonstante
6.670×10^{-11} N m²/kg²

GYROMAGNETIC RATIO γ Dim: $M^{-1}TI$

rapport gyromagnétique

gyromagnetisches Verhältnis

SI unit: ampere square metre per joule second A m^2/J s

H

HALF-LIFE $T_{\frac{1}{2}}$ Dim: T

période radioactive

Halbwertszeit

SI unit: second s

HALF-THICKNESS $d_{\frac{1}{2}}$ Dim: L

couche de demi-atténuation

Halbwertsschicht

SI unit: metre m

HEAT Q Dim: L^2MT^{-2}

quantité de chaleur

Wärme; Wärmemenge

SI unit: joule J

Note: also called quantity of heat; cf. work and energy

HEAT CAPACITY C Dim: $L^2MT^{-2}\Theta^{-1}$

capacité thermique

Wärmekapazität

SI unit: joule per kelvin J/K

other unit: joule per degree Celsius J/°C

HEAT FLOW RATE Φ, (q) Dim: L^2MT^{-3}

flux thermique

Wärmestrom

SI unit: watt W (kW)

Note: cf. power

HEAT TRANSFER COEFFICIENT see: coefficient of heat transfer

HEIGHT h Dim: L

hauteur

Höhe

SI unit: metre m

HELMHOLTZ FUNCTION

fonction de Helmholtz
Helmholtz-Funktion
Note: the same as free energy (q.v.)

I

ILLUMINATION E, (E_V) Dim: $L^{-2}J$

éclairement lumineux; éclairement
Beleuchtungsstärke
SI unit: lux lx
Note: also called illuminance

IMPEDANCE Z Dim: $L^2MT^{-3}I^{-2}$

impédance
Scheinwiderstand
SI unit: ohm Ω

INTEGRAL ABSORBED DOSE E_D Dim: L^2MT^{-2}

dose absorbée integrale
integrale Energiedosis
Note: other name for energy imparted (q.v.)

INTENSITY OF ELECTRIC CURRENT see: electric current

INTENSITY OF HEAT FLOW RATE see: density of heat flow rate

INTERNAL ENERGY U, (E) Dim: L^2MT^{-2}

énergie interne
innere Energie
SI unit: joule J

ION NUMBER DENSITY n^+, n^- Dim: L^{-3}

nombre volumique d'ions
Ionendichte
SI unit: reciprocal cubic metre m^{-3}
Note: also called ion density

IONIZATION EXPOSURE see: exposure

IONIZATION EXPOSURE RATE see: exposure rate

IRRADIANCE E, (E_e) Dim: MT^{-3}

éclairement énergétique

Bestrahlungsstärke

SI unit: watt per square metre W/m^2

K

KELVIN TEMPERATURE

obsol. name for thermodynamic temperature (q.v.)

KERMA K Dim: L^2T^{-2}

kerma

Kerma

SI unit: joule per kilogramme J/kg

Note: short for *k*inetic *e*nergy *r*eleased in *ma*tter; also called first collision dose

KERMA RATE $\dot{K}$ Dim: L^2T^{-3}

débit de kerma

Kermarate; Kermaleistung

SI unit: watt per kilogramme W/kg

KINEMATIC VISCOSITY ν Dim: L^2T^{-1}

viscosité cinématique

kinematische Viskosität

SI unit: metre squared per second m^2/s

other unit: centistokes cSt

KINETIC ENERGY E_k, K, T Dim: L^2MT^{-2}

énergie cinétique

kinetische Energie

SI unit: joule J

L

LATENT HEAT L Dim: L^2MT^{-2}

chaleur de transformation

latente Wärmemenge

SI unit: joule J

LATERAL CONTRACTION (η) Dim: 1

contraction latérale (relative)

Querkürzung

LENGTH l Dim: L

longueur

Länge

SI unit: metre m

LENGTH OF PATH s Dim: L

longueur curviligne; parcours

Weglänge

SI unit: metre m

LETHARGY u Dim: 1

léthargie

Lethargie

LEVEL WIDTH Γ Dim: L^2MT^{-2}

largeur de niveau

Breite des Resonanzniveau

SI unit: joule J

other unit: electronvolt eV

LIGHT EXPOSURE H Dim: $L^{-2}TJ$

exposition lumineuse

Belichtung

SI unit: lux second lx s

Note: formerly called quantity of illumination

LINEAR ATTENUATION COEFFICIENT μ, μ_l Dim: L^{-1}

coefficient linéique d'atténuation

linearer Schwächungskoeffizient

SI unit: reciprocal metre m^{-1}

LINEAR ENERGY TRANSFER L Dim: LMT^{-2}

transfert linéique d'énergie

lineares Energieübertragungsvermögen

SI unit: joule per metre J/m

other unit: electronvolt per centimetre eV/cm

LINEAR EXPANSION COEFFICIENT α, λ Dim: Θ^{-1}

coefficient de dilatation linéique

Längen-Ausdehnungskoeffizient

SI unit: reciprocal kelvin K^{-1}

other unit: reciprocal degree Celsius $°C^{-1}$

Note: also called thermal coefficient of linear expansion

LINEAR IONIZATION N_{il} Dim: L^{-1}

ionisation linéique

lineare Ionisation

SI unit: reciprocal metre m^{-1}

LINEAR STOPPING POWER S, S_l Dim: LMT^{-2}

pouvoir linéique d'arrêt

lineares Bremsvermögen

SI unit: joule per metre J/m

other unit: electronvolt per centimetre eV/cm

LINEAR STRAIN e, ε Dim: 1

dilatation linéique relative

Dehnung

Note: also called relative elongation

LOGARITHMIC DECREMENT Λ Dim: 1

décrément logarithmique

logarithmisches Dekrement

Note: neper Np is sometimes used as unit of this quantity

LOUDNESS N Dim: 1

sonie

Lautheit

unit: sone (S)

LOUDNESS LEVEL $L_N, (\Lambda)$ Dim: 1

niveau d'isosonie

Lautstärkepegel

unit: phon (Ph or P)

LUMINANCE $L, (L_V)$ Dim: $L^{-2}J$

luminance

Leuchdichte

SI unit: candela per square metre cd/m^2

Note: also called brightness

LUMINOUS EFFICACY K Dim: $L^{-2}M^{-1}T^3J$

efficacité lumineuse

photometrisches Strahlungsäquivalent

SI unit: lumen per watt lm/W

LUMINOUS EFFICIENCY V Dim: 1

efficacité lumineuse relative

Hellempfindlichkeitsgrad

LUMINOUS EMITTANCE see: luminous exitance

LUMINOUS ENERGY see: quantity of light

LUMINOUS EXITANCE M, (M_V) Dim: $L^{-2}J$

exitance lumineuse

spezifische Lichtausstrahlung

SI unit: lumen per square metre lm/m^2

Note: formerly called luminous emittance

LUMINOUS FLUX Φ, (Φ_V) Dim: J

flux lumineux

Lichstrom

SI unit: lumen lm

LUMINOUS INTENSITY I, (I_V) Dim: J

intensité lumineuse

Lichtstärke

SI unit: candela cd

Note: in this dictionary the symbol J is used when luminous intensity appears in the dimension of a quantity to avoid confusion with the symbol I used for electric current

M

MACROSCOPIC CROSS SECTION Σ Dim: L^{-1}

section efficace macroscopique

makroskopischer Wirkungsquerschnitt

SI unit: reciprocal metre m^{-1}

MAGNETIC DIPOLE MOMENT p Dim: $L^3MT^{-2}I^{-1}$

moment de dipôle magnétique

das Coulombsche magnetische Moment

SI unit: newton metre squared per ampere N m^2/A

or weber metre Wb m

MAGNETIC FIELD STRENGTH H; $\boldsymbol{H}$ Dim: $L^{-1}I$

champ magnétique
magnetische Feldstärke
SI unit: ampere per metre A/m

MAGNETIC FLUX Φ Dim: $L^2MT^{-2}I^{-1}$

flux magnétique; flux d'induction magnétique
magnetischer Fluss; magnetischer Induktionsfluss
SI unit: weber Wb
Note: also called flux of magnetic induction

MAGNETIC FLUX DENSITY B; $\boldsymbol{B}$ Dim: $MT^{-2}I^{-1}$

densité de flux magnétique; induction magnétique
magnetische Flussdichte; magnetische Induktion
SI unit: tesla T
Note: also called magnetic induction

MAGNETIC INDUCTION see: magnetic flux density

MAGNETIC MOMENT μ; $\boldsymbol{\mu}$ Dim: L^2I

moment magnétique
magnetisches Moment
SI unit: ampere square metre $A\ m^2$
Note: see electromagnetic moment

MAGNETIC POLARIZATION B_i, J; $\boldsymbol{J}$ Dim: $MT^{-2}I^{-1}$

polarisation magnétique
magnetische Polarisation
SI unit: tesla T

MAGNETIC POTENTIAL DIFFERENCE U_m Dim: I

différence de potential magnétique
magnetische Spannung
SI unit: ampere A

MAGNETIC SUSCEPTIBILITY κ; χ_m Dim: 1

susceptibilité magnétique
magnetische Suszeptibilität

MAGNETIC VECTOR POTENTIAL A; $\boldsymbol{A}$ Dim: $LMT^{-2}I^{-1}$

potentiel vecteur magnétique
magnetisches Vektorpotential
SI unit: weber per metre Wb/m

MAGNETIZATION H_i, M; $\mathbf{M}$ Dim: $L^{-1}I$

aimantation

Magnetisierung

SI unit: ampere per metre A/m

MAGNETOMOTIVE FORCE F, F_m Dim: I

force magnétomotrice

magnetomotorische Kraft

SI unit: ampere A

MASS m Dim: M

masse

Masse

SI unit: kilogramme kg

other unit: tonne t

MASS ATTENUATION COEFFICIENT μ/ρ, μ_m Dim: L^2M^{-1}

coefficient massique d'atténuation

Massenschwächungskoeffizient

SI unit: square metre per kilogramme m^2/kg

MASS CONCENTRATION OF COMPONENT B ρ_B Dim: $L^{-3}M$

concentration de masse du constituant B

Massenkonzentration der Komponente B

SI unit: kilogramme per cubic metre kg/m^3

MASS DEFECT B Dim: M

défaut de masse

Massendefekt

SI unit: kilogramme kg

other unit: (unified) atomic mass unit u

MASS ENERGY TRANSFER COEFFICIENT μ_{mk}, μ_K/ρ Dim: L^2M^{-1}

coefficient de transfert masse-énergie

Massen-Energieübertragungskoeffizient

SI unit: square metre per kilogramme m^2/kg

MASS EXCESS Δ Dim: M

excès de masse

Massenüberschuss

SI unit: kilogramme kg

other unit: (unified) atomic mass unit u

MASS NUMBER A Dim: 1

nombre de masse
Massenzahl
Note: also called nucleon number

MASS RATE OF (FLUID) FLOW $(q_m, \dot{M})$ Dim: MT^{-1}

débit en masse
Massenstrom
SI unit: kilogramme per second kg/s

MASS STOPPING POWER S/ρ, (S_m) Dim: L^4T^{-2}

pouvoir massique d'arrêt
Massenbremsvermögen
SI unit: joule square metre per kilogramme $J\ m^2/kg$
other unit: electronvolt square centimetre per gramme $eV\ cm^2/g$

MAXIMUM SPECTRAL LUMINOUS EFFICACY K_m Dim: $L^{-2}M^{-1}T^3J$

efficacité lumineuse spectrale maximale
Maximalwert des photometrischen Strahlungsäquivalents
SI unit: lumen per watt lm/W

MECHANICAL IMPEDANCE Z_m, (w) Dim: MT^{-1}

impédance mécanique
mechanische Impedanz
(SI) unit: newton second per metre N s/m

MECHANICAL STRESS see: stress

MEAN FREE PATH l, λ Dim: L

libre parcours moyen
mittlere freie Weglänge
SI unit: metre m

MEAN LIFE τ Dim: T

vie moyenne
mittlere Lebensdauer
SI unit: second s

MEAN LINEAR RANGE R, R_1 Dim: L

parcours moyen
mittlere (lineare) Reichweite
SI unit: metre m

MEAN MASS RANGE $R\rho$, (R_m) Dim: $L^{-2}M$

parcours moyen en masse

mittlere Massenreichweite

SI unit: kilogramme per square metre kg/m^2

MIGRATION AREA M^2 Dim: L^2

aire de migration

Wanderfläche

SI unit: square metre m^2

MIGRATION LENGTH M Dim: L

longueur de migration

Wanderlänge

SI unit: metre m

MOBILITY b, μ Dim: $M^{-1}T^2I$

mobilité

Beweglichkeit

SI unit: square metre per volt second $m^2/V\ s$

MODULUS OF COMPRESSION see: bulk modulus

MODULUS OF ELASTICITY see: Young's modulus

MODULUS OF RIGIDITY see: shear modulus

MODULUS OF SECTION Z, W Dim: L^3

module d'inertie

Widerstandsmoment

SI unit: metre cubed m^3

Note: also called section modulus

MOLALITY OF SOLUTE SUBSTANCE B m_B Dim: M^{-1}

molalité du constituant dissous B

Molalität der gelösten Substanz B

(SI) unit: mole per kilogramme mol/kg (kmol/kg)

MOLAR ENTROPY S_m Dim: $L^2MT^{-2}\Theta^{-1}$

entropie molaire

molare Entropie

(SI) unit: joule per mole kelvin J/mol K

MOLAR GAS CONSTANT R Dim: $L^2MT^{-2}\Theta^{-1}$

constante molaire des gaz

molare Gaskonstante

8.3143 J/mol K

MOLAR HEAT CAPACITY C_m Dim: $L^2MT^{-2}\Theta^{-1}$

capacité thermique molaire

molare Wärmekapazität

(SI) unit: joule per mole kelvin J/mol K

MOLAR INTERNAL ENERGY U_m; (E_m) Dim: L^2MT^{-2}

énergie interne molaire

molare innere Energie

(SI) unit: joule per mole J/mol

MOLAR MASS M Dim: M

masse molaire

molare Masse; stoffmengenbezogene Masse

(SI) unit: kilogramme per mole kg/mol

MOLAR VOLUME V_m Dim: L^3

volume molaire

molares Volumen

(SI) unit: cubic metre per mole m^3/mol

MOLARITY OF COMPONENT B c_B Dim: L^{-3}

molarité du constituant B;

concentration de quantité de matière du constituant B

Molarität der Komponente B;

Stoffmengenkonzentration der Komponente B

(SI) unit: mole per cubic metre mol/m^3

Note: also called amount of substance concentration of component B

MOLE FRACTION OF COMPONENT B x_B Dim: 1

fraction molaire du constituant B

Mollenbruch der Komponente B

MOLECULAR WEIGHT see: relative molecular mass

MOMENT OF A COUPLE see: torque

MOMENT OF FORCE M; $\boldsymbol{M}$ Dim: L^2MT^{-2}

moment d'une force

Moment einer Kraft

SI unit: newton metre N m

MOMENT OF INERTIA I, J Dim: L^2M

moment d'inertie

Massenträgheitsmoment

SI unit: kilogramme metre squared $kg\ m^2$

MOMENT OF MOMENTUM see: angular momentum

MOMENTUM $p; \boldsymbol{p}$ Dim: LMT^{-1}

quantité de mouvement

Impuls; (*Bewegungsgrösse*)

SI unit: kilogramme metre per second kg m/s

MULTIPLICATION FACTOR k Dim: 1

facteur de multiplication

Multiplikationsfaktor

MUTUAL INDUCTANCE M, L_{12} Dim: $L^2MT^{-2}I^{-2}$

inductance mutuelle

Gegeninduktivität

SI unit: henry H

N

NATURAL (NAPIERIAN) LOGARITHM OF THE RATIO OF TWO AMPLITUDES ($\log_e (A_1/A_2)$) Dim: 1

logarithme népérien (*naturel*) *du rapport de deux amplitudes*

natürlicher Logarithmus des Verhältnisses zweier Amplituden

unit: neper Np

NEUTRON FLUX DENSITY φ Dim: $L^{-2}T^{-1}$

densité de flux de neutrons

Neutronenflussdichte

SI unit: reciprocal second reciprocal square metre $s^{-1}\ m^{-2}$

NEUTRON NUMBER N Dim: 1

nombre de neutrons

Neutronenzahl

NEUTRON NUMBER DENSITY n Dim: L^{-3}

nombre volumique de neutrons

Neutronenzahldichte

SI unit: reciprocal cubic metre m^{-3}

NORMAL STRESS σ Dim: $L^{-1}MT^{-2}$

contrainte normale; tension normale

Normalspannung

SI unit: newton per square metre N/m^2

NUCLEAR MAGNETON μ_N Dim: L^2I

magnéton nucléaire

Kernmagneton

5.0505×10^{-27} J/T or A m^2

NUCLEON NUMBER A Dim: 1

nombre de nucléons

Nukleonzahl

Note: also called mass number

NUCLEAR QUADRUPOLE MOMENT Q Dim: L^2

moment quadripolaire nucléaire

Quadrupolmoment eines Kernes

SI unit: square metre m^2

NUCLEAR RADIUS R Dim: L

rayon nucléaire

Kernradius

SI unit: metre m

NUMBER OF MOLECULES (OR PARTICLES) N Dim: 1

nombre de molécules; nombre de particules

Molekülanzahl; Teilchenanzahl

NUMBER OF TURNS IN WINDING N Dim: 1

nombre de tours spires de l'enroulement

Windungszahl

P

PARTIAL PRESSURE OF COMPONENT B p_B Dim: $L^{-1}MT^{-2}$

pression partielle du constituant B

Partialdruck der Komponente B

SI unit: newton per square metre N/m^2

PARTICLE FLUX DENSITY φ Dim: $L^{-2}T^{-1}$

densité de flux de particules
Teilchenflussdichte
SI unit: reciprocal second reciprocal square metre
$s^{-1}\ m^{-2}$

PERIODIC TIME T Dim: T

durée d'une période
Periodendauer
SI unit: second s
Note: for other units see time

PERMEABILITY μ Dim: $LMT^{-2}I^{-2}$

perméabilité
Permeabilität
SI unit: henry per metre H/m

PERMEABILITY OF VACUUM μ_0 Dim: $LMT^{-2}I^{-2}$

perméabilité du vide
magnetische Feldkonstante; (*Induktionskonstante*)
$1.256\,64 \times 10^{-6}$ H/m

PERMEANCE Λ, (P) Dim: $L^2MT^{-2}I^{-2}$

perméance
magnetischer Leitwert
SI unit: henry H

PERMITTIVITY ε Dim: $L^{-3}M^{-1}T^4I^2$

permittivité
Permittivität; (*Dielektrizitätskonstante*)
SI unit: farad per metre F/m

PERMITTIVITY OF VACUUM ε_0 Dim: $L^{-3}M^{-1}T^4I^2$

permittivité du vide
elektrische Feldkonstante
SI unit: farad per metre F/m

PHASE COEFFICIENT β Dim: L^{-1}

constante de phase
Phasenkoeffizient
SI unit: reciprocal metre m^{-1}
Note: neper per metre Np/m is sometimes used as a unit of this quantity

PLANCK CONSTANT h Dim: L^2MT^{-1}

constante de Planck

Plancksches Wirkungsquantum; Plancksche Konstante

$6.625\,6 \times 10^{-34}$ J s

Note: $\hbar = h/2\pi \approx 1.054\,50 \times 10^{-34}$ J s

$h/e \approx 4.135\,56 \times 10^{-15}$ J s/C

$h/k \approx 4.799\,3 \times 10^{-11}$ s K

POISSON NUMBER see: Poisson ratio

POISSON RATIO u Dim: 1

coefficient de Poisson; nombre de Poisson; rapport de Poisson

Poisson-Zahl

Note: also called Poisson number

POTENTIAL DIFFERENCE $U, (V)$ Dim: $L^2MT^{-3}I^{-1}$

différence de potentiel; tension

elektrische Spannung

SI unit: volt V

Note: also called tension

POTENTIAL ENERGY E_p, U, V, Φ Dim: L^2MT^{-2}

énergie potentielle

potentielle Energie

SI unit: joule J

POWER P Dim: L^2MT^{-3}

puissance

Leistung

SI unit: watt W

POWER OF A LENS (D) Dim: L^{-1}

puissance d'une lentille; vergence

Brechkraft

(SI) unit: reciprocal metre m^{-1}

other unit: dioptre (dpt)

POYNTING VECTOR $S; \boldsymbol{S}$ Dim: MT^{-3}

vecteur de Poynting

Poynting-Vektor

SI unit: watt per square metre W/m^2

PRESSURE p Dim: $L^{-1}MT^{-2}$

pression

Druck

SI unit: newton per square metre N/m^2

other unit: bar bar

PRESSURE COEFFICIENT β Dim: Θ^{-1}

coefficient d'augmentation de pression

Spannungskoeffizient

SI unit: reciprocal kelvin K^{-1}

other unit: reciprocal degree Celsius $°C^{-1}$

PROPAGATION COEFFICIENT γ Dim: L^{-1}

constante de propagation

Fortpflanzungskonstante

Ausbreitungskoeffizient

SI unit: reciprocal metre m^{-1}

Note: neper per metre Np/m is sometimes used as a unit of this quantity

PROTON NUMBER Z Dim: 1

nombre de protons

Protonenzahl

Note: also called atomic number

PULSATANCE see: angular frequency

Q

QUALITY FACTOR Q Dim: 1

facteur de qualité

Gütefaktor

QUANTITY OF ELECTRICITY see: electric charge

QUANTITY OF HEAT see: heat

QUANTITY OF ILLUMINATION see: light exposure

QUANTITY OF LIGHT $Q, (Q_V)$ Dim: TJ

quantité de lumière

Lichtmenge

SI unit: lumen second lm s

Note: also called luminous energy

R

RADIANCE L, (L_e) Dim: MT^{-3}

luminance énergétique; radiance

Strahldichte

SI unit: watt per steradian square metre $W/sr\ m^2$

RADIANT ENERGY Q, W, (U, Q_e) Dim: L^2MT^{-2}

énergie rayonnante

Strahlungsenergie; Strahlungsmenge

SI unit: joule J

RADIANT EXITANCE M, (M_e) Dim: MT^{-3}

exitance énergétique

spezifische Ausstrahlung

SI unit: watt per square metre W/m^2

Note: formerly called radiant emittance

RADIANT FLUX P, Φ, (Φ_e) Dim: L^2MT^{-3}

flux énergétique; puissance rayonnante

Strahlungsfluss

SI unit: watt W

Note: also called radiant power

RADIANT INTENSITY I, (I_e) Dim: L^2MT^{-3}

intensité énergétique

Strahlstärke

SI unit: watt per steradian W/sr

RADIUS r Dim: L

rayon

Radius; Halbmesser

SI unit: metre m

RATIO OF SPECIFIC HEAT CAPACITIES γ, κ Dim: 1

rapport des chaleurs massiques

Verhältnis der spezifischen Wärmekapazitäten

REACTANCE X Dim: $L^2MT^{-3}I^{-2}$

réactance

Blindwiderstand

SI unit: ohm Ω

REACTIVE POWER Q, (P_q) Dim: L^2MT^{-3}

puissance réactive

Blindleistung

(SI) unit: var var

REACTIVITY ρ Dim: 1

réactivité

Reaktivität

REACTOR TIME CONSTANT T Dim: T

constante de temps du réacteur; période du réacteur

Reaktorzeitkonstante; Reaktorperiode

SI unit: second s

Note: also called reactor period

RECOMBINATION COEFFICIENT α Dim: L^3T^{-1}

coefficient de recombinaison

Rekombinationskoeffizient

SI unit: cubic metre per second m^3/s

REFLECTION COEFFICIENT r, ρ Dim: 1

facteur de réflection

Schallreflexionsgrad

REFRACTIVE INDEX n Dim: 1

indice de réfraction

Brechzahl

RELATIVE ATOMIC MASS OF AN ELEMENT A_r Dim: 1

masse atomique relative d'un élément

relative Atommasse eines Elementes

Note: formerly called atomic weight

RELATIVE DENSITY d Dim: 1

densité relative

relative Dichte

Note: cf. specific gravity

RELATIVE ELONGATION see: linear strain

RELATIVE MOLECULAR MASS OF A SUBSTANCE M_r Dim: 1

masse moléculaire relative d'une substance

relative Molekülmasse eines Stoffes

Note: formerly called molecular weight

RELATIVE PERMEABILITY μ_r Dim: 1

perméabilité relative

Permeabilitätszahl; (relative Permeabilität)

RELATIVE PERMITTIVITY ε_r Dim: 1

permittivité relative

Permittivitätszahl; (relative Permittivität)

RELUCTANCE R, R_m Dim: $L^{-2}M^{-1}T^2I^2$

réluctance

magnetischer Widerstand

SI unit: reciprocal henry H^{-1}

RESISTANCE R Dim: $L^2MT^{-3}I^{-2}$

résistance

elektrischer Widerstand

SI unit: ohm Ω

Note: reciprocal of resistance is called conductance

RESISTIVITY ρ Dim: $L^3MT^{-3}I^{-2}$

résistivité

spezifischer elektrischer Widerstand

SI unit: ohm metre Ω m

Note: reciprocal of resistivity is called conductivity

RESONANCE ENERGY E_r, E_{res} Dim: L^2MT^{-2}

énergie de résonance

Resonanzenergie

SI unit: joule J

other unit: electronvolt eV

REST MASS OF ELECTRON m_e Dim: M

masse au repos de l'électron

Ruhmasse des Elektrons

9.1091×10^{-31} kg

$5.485\,97 \times 10^{-4}$ u

REST MASS OF NEUTRON m_n Dim: M

masse au repos du neutron

Ruhmasse des Neutrons

$1.674\,82 \times 10^{-27}$ kg

1.008 67 u

REST MASS OF PROTON m_p Dim: M

masse au repos du proton

Ruhmasse des Protons

$1.672\,52 \times 10^{-27}$ kg

1.007 28 u

Note: m_p/m_e $1.836\,10 \times 10^3$

REVERBERATION TIME T Dim: T

duré de réverbération

Nachhallzeit

SI unit: second s

ROTATIONAL FREQUENCY n Dim: T^{-1}

fréquence de rotation

Drehfrequenz; (Drehzahl)

SI unit: reciprocal second s^{-1}

other units: revolution per minute (rev/min)

revolution per second (rev/s)

Note: cf. angular velocity

RYDBERG CONSTANT R Dim: L^{-1}

constante de Rydberg

Rydberg Konstante

$1.097\,37 \times 10^7$ m^{-1}

Note: $R_H \approx 1.096\,78 \times 10^7$ m^{-1}

$R_\infty c \approx 3.289\,84 \times 10^{-15}$ s^{-1}

$R_\infty hc \approx 2.179\,72 \times 10^{-18}$ J

S

SECOND (AXIAL) MOMENT OF AREA I, I_a Dim: L^4

moment quadratique (axial) d'une aire plane

(axiales) Flächenträgheitsmoment

SI unit: metre to the fourth m^4

SECOND POLAR MOMENT OF AREA I_p, J Dim: L^4

moment quadratique polaire d'une aire plane

polares Flächenträgheitsmoment

SI unit: metre to the fourth m^4

SECOND RADIATION CONSTANT c_2 Dim: LΘ
seconde constante de rayonnement
die zweite Strahlungskonstante
$1.438\,79 \times 10^{-2}$ m K

SECTION MODULUS see: modulus of section

SELF INDUCTANCE L Dim: $L^2MT^{-2}I^{-2}$
inductance propre
Selbstinduktivität; Eigeninduktivität
SI unit: henry H

SHEAR ANGLE see: shear strain

SHEAR MODULUS G Dim: $L^{-1}MT^{-2}$
module d'elasticité de glissement; module de torsion; module de Coulomb
Schubmodul
SI unit: newton per square metre N/m^2
Note: also called modulus of rigidity

SHEAR STRAIN γ Dim: 1
glissement unitaire
Schiebung
Note: also called shear angle

SHEAR STRESS τ Dim: $L^{-1}MT^{-2}$
contrainte tangentielle; tension de cisaillement
Schubspannung; Scherspannung
SI unit: newton per square metre N/m^2

SLOWING DOWN AREA L_s^2, L_{s1}^2 Dim: L^2
aire de ralentissement
Bremsfläche
SI unit: square metre m^2

SLOWING DOWN LENGTH L_s, L_{s1} Dim: L
longueur de ralentissement
Bremslänge
SI unit: metre m

SOLID ANGLE Ω, ω Dim: 1
angle solide
Raumwinkel; räumlicher Winkel
SI unit: steradian sr

SOUND ENERGY DENSITY E Dim: $L^{-1}MT^{-2}$

énergie volumique acoustique
Schallenergiedichte
SI unit: joule per cubic metre J/m^3

SOUND ENERGY FLUX $P, (N, W)$ Dim: L^2MT^{-3}

flux d'énergie acoustique; puissance acoustique
Schalleistung
(SI) unit: watt W
Note: also called sound power or acoustic power

SOUND INTENSITY I, J Dim: MT^{-3}

intensité acoustique
Schallintensität
(SI) unit: watt per square metre W/m^2

SOUND INTENSITY LEVEL I, J Dim: 1

niveau d'intensité acoustique
Schallintensitätspegel
unit: decibel dB

SOUND PARTICLE ACCELERATION a Dim: LT^{-2}

accélération acoustique d'une particule
Schallteilchenbeschleunigung
SI unit: metre per second squared m/s^2

SOUND PARTICLE DISPLACEMENT $\xi, (x)$ Dim: L

élongation
Schallausschlag
SI unit: metre m

SOUND PARTICLE VELOCITY u, v Dim: LT^{-1}

vitesse acoustique d'une particule
Schallschnelle
SI unit: metre per second m/s

SOUND POWER see: sound energy flux

SOUND POWER LEVEL $L_P, (L_N, L_W)$ Dim: 1

niveau de puissance acoustique
Schalleistungspegel
unit: decibel dB

SOUND PRESSURE p Dim: $L^{-1}MT^{-2}$

pression acoustique

Schalldruck

SI unit: newton per square metre N/m^2

SOUND PRESSURE LEVEL L_p, (L) Dim: 1

niveau de pression acoustique

Schalldruckpegel

unit: decibel dB

SOUND REDUCTION INDEX R Dim: 1

indice de réduction acoustique

Schalldämm-Mass

unit: decibel dB

Note: also called sound transmission loss

SOUND TRANSMISSION LOSS see: sound reduction index

SPECIFIC ACOUSTIC IMPEDANCE Z_s, (W) Dim: $L^{-2}MT^{-1}$

impédance acoustique spécifique

spezifische Schallimpedanz; Feldimpedanz

(SI) unit: newton second per metre cubed $N\ s/m^3$

SPECIFIC ACTIVITY a Dim: $M^{-1}T^{-1}$

activité massique; activité spécifique

spezifische Aktivität

SI unit: reciprocal second reciprocal kilogramme

$s^{-1}\ kg^{-1}$

other unit: curie per gramme Ci/g

SPECIFIC ENTHALPY h, (i) Dim: L^2T^{-2}

enthalpie massique

spezifische Enthalpie

SI unit: joule per kilogramme J/kg

SPECIFIC ENTROPY s Dim: $L^2T^{-2}\Theta^{-1}$

entropie massique

spezifische Entropie

SI unit: joule per kilogramme kelvin J/kg K

SPECIFIC FREE ENERGY f Dim: L^2T^{-2}

énergie libre massique

spezifische freie Energie

SI unit: joule per kilogramme J/kg

SPECIFIC FREE ENTHALPY see: specific Gibbs function

SPECIFIC GAMMA RAY CONSTANT Γ Dim: $L^2M^{-1}TI$

constante spécifique de rayonnement gamma
spezifische Gammastrahlenkonstante
SI unit: coulomb metre squared per kilogramme
$C\ m^2/kg$
other unit: röntgen metre squared per curie hour
$R\ m^2/Ci\ h$

SPECIFIC GIBBS FUNCTION g Dim: L^2T^{-2}

enthalpie libre massique
spezifische freie Enthalpie
SI unit: joule per kilogramme J/kg

SPECIFIC HEAT CAPACITY c Dim: $L^2T^{-2}\Theta^{-1}$

chaleur massique; (capacité thermique massique)
spezifische Wärmekapazität
SI unit: joule per kilogramme kelvin J/kg K
other unit: joule per kilogramme degree Celsius J/kg °C

SPECIFIC HEAT CAPACITY AT CONSTANT PRESSURE c_p Dim: $L^2T^{-2}\Theta^{-1}$

chaleur massique à pression constante
isobare spezifische Wärmekapazität;
spezifische Wärmekapazität bei konstantem Druck
SI unit: joule per kilogramme kelvin J/kg K
other unit: joule per kilogramme degree Celsius
J/kg °C

SPECIFIC HEAT CAPACITY AT CONSTANT VOLUME c_v Dim: $L^2T^{-2}\Theta^{-1}$

chaleur massique à volume constant
isochore spezifische Wärmekapazität;
spezifische Wärmekapazität bei konstantem Volumen
SI unit: joule per kilogramme kelvin J/kg K
other unit: joule per kilogramme degree Celsius
J/kg °C

SPECIFIC INTERNAL ENERGY u, (e) Dim: L^2T^{-2}

énergie interne massique

spezifische innere Energie

SI unit: joule per kilogramme J/kg

SPECIFIC LATENT HEAT l Dim: L^2T^{-2}

chaleur de transformation massique

spezifische latente Wärmemenge

SI unit: joule per kilogramme J/kg

SPECIFIC VOLUME v Dim: L^3M^{-1}

volume massique

spezifisches Volumen

SI unit: cubic metre per kilogramme m^3/kg

Note: reciprocal of specific volume is called density

SPECIFIC WEIGHT γ Dim: $L^{-2}MT^{-2}$

poids volumique; poids spécifique

Wichte; (spezifisches Gewicht)

SI unit: newton per cubic metre N/m^3

Note: also called weight density; this quantity should be avoided and mass density used instead

SPECTRAL ABSORPTANCE $\alpha(\lambda)$ Dim: 1

facteur spectral d'absorption; (absorptance spectrale)

spektraler Absorptionsgrad

Note: also called spectral absorption factor

SPECTRAL CROSS SECTION σ_E Dim: $M^{-1}T^2$

section efficace spectrale

spektraler Wirkungsquerschnitt

SI unit: square metre per joule m^2/J

other unit: barn per erg b/erg

SPECTRAL DIFFERENTIAL CROSS SECTION $\sigma_{\Omega,E}$ Dim: $M^{-1}T^2$

section efficace différentielle spectrale

spektraler differentieller Wirkungsquerschnitt

SI unit: square metre per steradian joule $m^2/sr\ J$

other unit: barn per steradian erg b/sr erg

SPECTRAL EMISSIVITY $\varepsilon(\lambda)$ Dim: 1

émissivité spectrale
spektraler Emissionsgrad

SPECTRAL LUMINOUS EFFICACY $K(\lambda)$ Dim: $L^{-2}M^{-1}T^{3}J$

efficacité lumineuse spectrale
spektrales photometrisches Strahlungsäquivalent
SI unit: lumen per watt lm/W

SPECTRAL LUMINOUS EFFICIENCY $V(\lambda)$ Dim: 1

efficacité lumineuse relative spectrale
spektraler Hellempfindlichkeitsgrad

SPECTRAL REFLECTANCE $\rho(\lambda)$ Dim: 1

facteur spectral de réflexion; (*réflectance spectrale*)
spektraler Reflexionsgrad
Note: also called spectral reflection factor

PECTRAL TRANSMITTANCE $\tau(\lambda)$ Dim: 1

facteur spectral de transmission; (*transmittance spectrale*)
spektraler Transmissionsgrad
Note: also called spectral transmission factor

SPEED see: velocity

STANDARD ACCELERATION see: standard gravity

STANDARD GRAVITY g_n Dim: LT^{-2}

valeur normale de l'accélération de la pesanteur;
accélération de la pesanteur normale
Normwert der Fallbeschleunigung; Normfallbeschleunigung
9.806 65 $m/s^2 \approx 32.1740\ ft/s^2$

STATIC PRESSURE p_s Dim: $L^{-1}MT^{-2}$

pression statique
statischer Druck
SI unit: newton per square metre N/m^2

STEFAN-BOLTZMANN CONSTANT σ Dim: $MT^{-3}\Theta^{-4}$

constante de Stefan-Boltzmann
Stefan-Boltzmann-Konstante
$5.6697 \times 10^{-8}\ W/m^2\ K^4$

STOICHIOMETRIC NUMBER ν Dim: 1

nombre stoechiométrique
stöchiometrische Zahl

STRESS σ Dim: $L^{-1}MT^{-2}$

contrainte; tension
Spannung
SI unit: newton per square metre N/m^2
Note: also called mechanical stress; cf. normal stress, shear stress, pressure

SURFACE DENSITY OF CHARGE σ Dim: $L^{-2}TI$

charge surfacique
Flächenladungsdichte
SI unit: coulomb per square metre C/m^2

SURFACE TENSION $\sigma, (\gamma)$ Dim: MT^{-2}

tension superficielle
Oberflächenspannung
SI unit: newton per metre N/m

SUSCEPTANCE B Dim: $L^{-2}M^{-1}T^{3}I^{2}$

susceptance
Blindleitwert
SI unit: reciprocal ohm Ω^{-1}
(SI) unit: siemens S

T

TEMPERATURE CONDUCTIVITY see: thermal diffusivity

TEMPERATURE DIFFERENCE see: temperature interval

TEMPERATURE INTERVAL $\Delta t, \Delta T$ Dim: Θ

intervalle de température; différence de température
Temperaturintervall; Temperaturdifferenz
SI unit: kelvin K
other unit: degree Celsius °C
Note: the commonly used abbreviation deg for temperature interval is now regarded as obsolescent and should be avoided

TEN TIMES THE COMMON (BRIGGSIAN) LOGARITHM OF THE RATIO OF TWO POWERS (OR ENERGIES) $10 \log_{10} (P_1/P_2)$ Dim: 1

dix fois le logarithme décimal (vulgaire) du rapport de deux puissances (ou de deux énergies)
zehnfacher Zehnerlogarithmus (Briggscher Logarithmus) des Verhältnisses zweier Leistungen (oder Energien)
unit: decibel dB

THERMAL COEFFICIENT OF LINEAR EXPANSION see: linear expansion coefficient

THERMAL CONDUCTANCE see: coefficient of heat transfer

THERMAL CONDUCTIVITY λ, (k) Dim: $LMT^{-3}\Theta^{-1}$

conductivité thermique
Wärmeleitfähigkeit
SI unit: watt per metre kelvin W/m K
other unit: watt per metre degree Celsius W/m °C
Note: reciprocal of thermal conductivity is called thermal resistivity; cf. thermal conductance

THERMAL DIFFUSIVITY a, (α, κ) Dim: L^2T^{-1}

diffusivité thermique
Temperaturleitfähigkeit
SI unit: metre squared per second m^2/s

THERMODYNAMIC TEMPERATURE T, Θ Dim: Θ

température thermodynamique
thermodynamische Temperatur
SI unit: kelvin K
Note: cf. Celsius temperature and temperature interval

THICKNESS d, δ Dim: L

épaisseur
Dicke
SI unit: metre m

TIME see: periodic time

TIME t Dim: T

temps
Zeit
SI unit: second s
other units: day d, hour h, minute min

TIME CONSTANT (of an exponentially varying quantity) τ, (T) Dim: T

constante de temps (d'une grandeur variant exponentiellement)

Zeitkonstante (einer exponentiell sich ändernden Grösse)

SI unit: second s

TORQUE *T; **T*** Dim: L^2MT^{-2}

torque; moment d'un couple

Drehmoment; Moment eines Kräftepaares

SI unit: newton metre N m

Note: also called moment of a couple

TOTAL CROSS SECTION σ_{tot}, σT Dim: L^{-1}

section efficace totale

totaler Wirkungsquerschnitt

SI unit: square metre m^2

other unit: barn b

TRANSMISSION COEFFICIENT τ Dim: 1

facteur de transmission

Schalltransmissionsgrad

V

VELOCITY see: angular velocity

VELOCITY (LINEAR VELOCITY) u, v, w, c Dim: LT^{-1}

vitesse; (vitesse linéaire)

Geschwindigkeit

SI unit: metre per second m/s

other units: kilometre per hour km/h

VELOCITY OF LIGHT IN VACUO c Dim: LT^{-1}

vitesse de la lumière dans le vide

Lichtgeschwindigkeit im Vakuum

$2.997\,925 \times 10^8$ m/s

Note: also called speed of light in vacuum

VELOCITY OF ROTATION see: angular velocity and rotational frequency

VELOCITY OF SOUND c Dim: LT^{-1}

vitesse du son

Schallgeschwindigkeit

SI unit: metre per second m/s

Note: see Mach number

VISCOSITY; (DYNAMIC VISCOSITY) η, (μ) Dim: $L^{-1}MT^{-1}$

viscosité; (viscosité dynamique)

Viskosität; (dynamische Viskosität)

SI unit: newton second per metre squared N s/m²

other unit: centipoise cP

Note: reciprocal of dynamic viscosity is called fluidity

VOLUME V, (v) Dim: L^3

volume

Volumen

SI unit: cubic metre m^3

other unit: litre l

VOLUME DENSITY OF CHARGE ρ Dim: $L^{-3}TI$

charge volumique

Raumladungsdichte

SI unit: coulomb per cubic metre C/m^3

Note: called charge density if no danger of mistake

VOLUME EXPANSION COEFFICIENT see: cubic expansion coefficient

VOLUME PER DISTANCE see: fuel consumption

VOLUME RATE OF (FLUID) FLOW (q_v, $\dot{V}$) Dim: L^3T^{-1}

débit en volume; débit-volume

Volumenstrom

SI unit: cubic metre per second m^3/s

VOLUME STRAIN θ, ϑ Dim: 1

dilatation volumique relative

relative Volumenänderung

Note: also called bulk strain

VOLUME VELOCITY *q*, *U* Dim: L^3T^{-1}

flux de vitesse acoustique

Schallfluss

SI unit: cubic metre per second m^3/s

W

WAVELENGTH λ Dim: L

longueur d'onde

Wellenlänge

SI unit: metre m

WAVE NUMBER σ, $(\tilde{\nu})$ Dim: L^{-1}

nombre d'onde linéique

Wellenzahl

SI unit: reciprocal metre m^{-1}

WEIGHT *G* (*P*, *W*) Dim: LMT^{-2}

poids

Gewichtskraft; Gewicht

SI unit: newton N

Note: this quantity should be avoided and mass used instead

WORK *A*, *W* Dim: L^2MT^{-2}

travail

Arbeit

SI unit: joule J

other units: kilowatt hour kwh; electronvolt eV

Y

YOUNG'S MODULUS *E* Dim: $L^{-1}MT^{-2}$

module d'élasticité longitudinale; module de Young

Elastizitätsmodul

SI unit: newton per square metre N/m^2

Note: also called modulus of elasticity

PART III

Symbols Denoting Quantities, Constants and Dimensions

SYMBOLS DENOTING QUANTITIES, CONSTANTS AND DIMENSIONS

A	activity, affinity, area, equivalent absorption area, magnetic vector potential, mass number, nucleon number, work
A_r	relative atomic mass of an element
a	acceleration (linear), sound particle acceleration, specific activity, thermal diffusivity
a_0	Bohr radius
B	magnetic flux density, mass defect, susceptance
B_i	magnetic polarization
b	breadth, mobility, moment of momentum
C	capitance, heat capacity
C_B	concentration of molecules of component B
C_m	molar heat capacity
c	velocity, specific heat capacity, velocity of light in vacuo, velocity of sound
c_B	molarity of component B
c_p	specific heat capacity at constant pressure
c_v	specific heat capacity at constant volume
c_1	first radiation constant
c_2	second radiation constant
D	absorbed dose, diffusion coefficient, (diffusion coefficient for neutron flux density), electric displacement, power of a lens
$\dot{D}$	absorbed dose rate
D_φ	diffusion coefficient for neutron flux density
d	diameter, relative density, thickness
$d_{\frac{1}{2}}$	half-thickness
E	electric field strength, electromotive force, energy, illumination, (internal energy), irradiance, sound energy density, Young's modulus
E_D	energy imparted, integral absorbed dose
E_e	(irradiance)
E_k	kinetic energy
E_m	(molar internal energy)
E_p	potential energy
E_r	resonance energy

E_{res}	resonance energy
E_v	(illumination)
e	elementary charge, linear strain, (specific internal energy)
F	Faraday constant, force, free energy, magnetomotive force
F_m	magnetomotive force
f	(coefficient of friction), focus length, frequency, specific free energy
G	conductance, Gibbs function, gravitational constant, shear modulus, weight
g	acceleration of free fall, specific Gibbs function
g_n	standard gravity
H	action, enthalpy, light exposure, magnetic field strength, magnetization
h	coefficient of heat transfer, height, Planck constant, specific enthalpy
I	electric current (also in dimension), luminous intensity, moment of inertia, radiant intensity, second moment of area, sound intensity, sound intensity level
I_a	second axial moment of area
I_e	(radiant intensity)
I_p	second polar moment of area
I_v	(luminous intensity)
i	frequency interval, (specific enthalpy)
J	(in dimension) luminous intensity, current density of particles, electric current density, magnetic polarization, moment of inertia, sound intensity, sound intensity level, second moment of area
K	bulk modulus, coefficient of heat transfer, (electric field strength), kerma, kinetic energy, luminous efficacy
$\dot{K}$	kerma rate
K_m	maximum spectral luminous efficacy
$K(\lambda)$	spectral luminous efficacy
k	Boltzmann constant, circular wave number, multiplication factor, (thermal conductivity)

L	(in dimensions) length, Avogadro constant, diffusion length, latent heat, linear energy transfer, luminance, radiance, self inductance
L_e	(radiance)
L_N	loudness level, (sound power level)
L_p	sound power level, sound pressure level
L_s	slowing down length
L_s^2	slowing down area
L_{sl}	slowing down length
L_{sl}^2	slowing down area
L_v	(luminance)
L_w	(sound power level)
L^2	diffusion area
L_{12}	mutual inductance
l	length, mean free path, specific latent heat
M	(in dimension) mass, bending moment, luminous exitance, magnetization, migration length, molar mass, moment of force, mutual inductance, radiant exitance
$\dot{M}$	mass rate of flow
M_e	(radiant exitance)
M_r	relative molecular mass of a substance
M_v	(luminous exitance)
M^2	migration area
m	electromagnetic moment, mass
$\dot{m}$	mass rate of flow
m_B	molality of solute substance B
m_e	rest mass of electron
m_n	rest mass of neutron
m_p	rest mass of proton
m_u	atomic mass constant (unified)
N	loudness, neutron number, number of molecules (or particles), number of turns in winding, (sound energy flux)
N_A	Avogadro constant
N_{il}	linear ionization
n	amount of substance, neutron number density, refractive index, rotational frequency

n_B	concentration of molecules of component B
n^+	ion number density
n^-	ion number density
P	active power, electric polarization, (permanence), power, radiant flux, sound energy flux, (weight)
P_q	(reactive power)
P_s	(apparent power)
p	electric dipole moment, magnetic dipole moment, momentum, pressure, sound pressure
p_B	partial pressure of component B
p_e	(electric dipole moment)
p_s	static pressure
p_ϑ	moment of momentum
Q	disintegration energy, electric charge, heat, nuclear quadrupole moment, quality factor, quantity of light, radiant energy, reactive power, (volume rate of flow)
Q_e	(radiant energy)
Q_v	(quantity of light)
q	density of heat flow rate, (heat flow rate), volume velocity, (mass and volume) rate of flow
R	mean linear range, molar gas constant, nuclear radius, reluctance, resistance, sound reduction index
R_l	mean linear range
R_m	reluctance, (mean mass range)
R_ρ	mean mass range
R_∞	Rydberg constant
r	radius, reflection coefficient
r_e	electron radius
S	apparent power, (area), (current density), (current density of particles), entropy, linear stopping power, Poynting vector
S_a	atomic stopping power
S_l	linear stopping power
S_m	(mass stopping power), molar entropy
S/ρ	mass stopping power
s	length of path, specific entropy

T	(in dimension) time, kinetic energy, periodic time, reactor time constant, reverberation time, thermodynamic temperature, (time constant), torque
$T_{\frac{1}{2}}$	half-title
t	Celsius temperature, time
U	coefficient of heat transfer, internal energy, potential difference, potential energy, (radiant energy), volume velocity
U_{m}	magnetic potential difference, molar internal energy
u	lethargy, sound particle velocity, specific internal energy, velocity
V	electric potential, luminous efficiency, (potential difference), potential energy, volume
$\dot{V}$	volume rate of flow
V_{m}	molar volume
$V(\lambda)$	spectral luminous efficiency
v	sound particle velocity, specific volume, velocity, (volume)
W	energy, modulus of section, radiant energy, (sound energy flux), (specific acoustic impedance), (weight), work
w	electromagnetic energy density, (mechanical impedance), velocity
X	exposure, reactance
$\dot{X}$	exposure rate
x	(sound particle displacement)
x_{B}	mole fraction of component B
Y	admittance
Z	(acoustic impedance), atomic number, impedance, modulus of section, proton number
Z_{a}	acoustic impedance
Z_{m}	mechanical impedance
Z_{s}	specific acoustic impedance
α	accoustic absorption coefficient, angle (plane), angular acceleration, attenuation coefficient, coefficient of heat transfer, cubic expansion coefficient, fine structure constant, linear expansion coefficient, recombination coefficient, (thermal diffusivity)

α_a	(acoustic absorption coefficient)
$\alpha(\lambda)$	spectral absorptance
β	phase coefficient, angle (plane), cubic expansion coefficient, pressure coefficient
Γ	level width, specific gamma ray constant
γ	conductivity, cubic expansion coefficient, gyromagnetic ratio, angle (plane), propagation coefficient, ratio of specific heat capacities, shear strain, specific weight, (surface tension)
Δ	mass excess
δ	damping coefficient, dissipation coefficient, thickness
ε	emissivity, fast fission factor, linear strain, permittivity
ε_0	permittivity of vacuum
ε_r	relative permittivity
$\varepsilon(\lambda)$	spectral emissivity
$\varepsilon(\lambda, \theta)$	directional spectral emissivity
η	efficiency, linear contraction, viscosity (dynamic)
Θ	(in dimension) temperature, thermodynamic temperature
θ, ϑ	angle (plane), Celsius temperature, (directional spectral emissivity), volume strain
κ	compressibility, magnetic susceptibility, (thermal diffusivity)
Λ	logarithmic decrement, (loudness level), permeance
λ	decay constant, (directional spectral emissivity), linear expansion coefficient, mean free path, thermal conductivity, wavelength
λ_C	Compton wavelength
μ	coefficient of friction, (dynamic viscosity), linear attenuation coefficient, magnetic moment, mobility, permeability, Poisson's ratio
μ_a	atomic attenuation coefficient
μ_{at}	atomic attenuation coefficient
μ_B	Bohr magneton, chemical potential of component B
μ_e	linear attenuation coefficient
μ_K/ρ	mass energy transfer coefficient

μ_{m}	mass attenuation coefficient
μ_{mk}	mass energy transfer coefficient
μ_{N}	nuclear magneton
μ_0	permeability of vacuum
μ_{r}	relative permeability
μ/ρ	mass attenuation coefficient
ν	(amount of substance), frequency, kinematic viscosity, Poisson ratio, stoichiometric number
$\tilde{\nu}$	(wave number)
ξ	average logarithmic energy decrement, sound particle displacement
ρ	density, reactivity, reflection coefficient, resistivity, volume density of charge
ρ_{B}	mass concentration of component B
$\rho(\lambda)$	spectral reflectance
Σ	macroscopic cross section
σ	conductivity, cross section, normal stress, Stefan-Boltzmann constant, stress, surface density of charge, surface tension, wave number
σ_{E}	spectral cross section
σ_{T}	total cross section
σ_{tot}	total cross section
σ_{Ω}	differential cross section
$\sigma_{\Omega,\mathrm{E}}$	spectral differential cross section
τ	mean life, shear stress, time constant, transmission coefficient, spectral transmittance
Φ	heat flow rate, luminous flux, magnetic flux, potential energy, radiant flux
Φ_{e}	(radiant flux)
Φ_{v}	(luminous flux)
φ	angle (plane), (density of heat flow rate), (directional spectral emissivity), electric potential, fluidity, neutron flux density, particle flux density
χ_{e}	electric susceptibility
Ψ	electric flux
ψ	energy flux density
Ω	solid angle
ω	angular frequency, angular velocity, solid angle

Appendices

(a) UK AND US UNITS

Table 1 UK and US units of length

	in	ft	yd	chain	furlong	mile
in	1	1/12	1/36	1/792	1/7920	1/63 360
ft	12	1	1/3	1/66	1/660	1/5280
yd	36	3	1	1/22	1/220	1/1760
chain	792	66	22	1	1/10	1/80
furlong	7920	660	220	10	1	1/8
mile	63 360	5280	1760	80	8	1

Table 2 UK and US units of area

	in^2	ft^2	yd^2	rood	acre	$mile^2$
in^2	1	1/144	1/1296	—	—	—
ft^2	144	1	1/9	1/10 890	1/43 560	—
yd^2	1296	9	1	1/1210	1/4840	$3.228\,31 \times 10^{-7}$
rood	—	10 890	1210	1	1/4	1/2560
acre	—	43 560	4840	4	1	1/640
$mile^2$	—	—	$3.097\,6 \times 10^{6}$	2560	640	1

Table 3 UK and US units of volume

	in^3	ft^3	yd^3
in^3	1	1/1728	1/46 656
ft^3	1728	1	1/27
yd^3	46 656	27	1
$mile^3$	$2.543\,58 \times 10^{14}$	$1.471\,98 \times 10^{11}$	$5.451\,78 \times 10^{9}$

Table 4 UK units of volume (capacity)

bushel	peck	gallon	quart	pint	gill	fl. oz	fl. dr	minim
1	4	8	32	64	256	1280	10240	614400
1/4	1	2	8	16	64	320	2560	153600
1/8	1/2	1	4	8	32	160	1280	76800
1/32	1/8	1/4	1	2	8	40	320	19200
1/64	1/16	1/8	1/2	1	4	20	160	9600
1/256	1/64	1/32	1/8	1/4	1	5	40	2400
1/1280	1/320	1/160	1/40	1/20	1/5	1	8	480
1/10240	1/2560	1/1280	1/320	1/160	1/40	1/8	1	60
1/614400	1/153600	1/76800	1/19200	1/9600	1/2400	1/480	1/60	1

Table 5 US units of volume (capacity)—for liquid measure only

	gallon	lq. qt.	lq. pt.	gill	fl. oz	fl. dr.	minim
gallon	1	4	8	32	128	1024	61440
lq. quart	1/4	1	2	8	32	256	15360
lq. pint	1/8	1/2	1	4	16	128	7680
gill	1/32	1/8	1/4	1	4	32	1920
fl. oz	1/128	1/32	1/16	1/4	1	8	480
fl. dr	1/1024	1/256	1/128	1/32	1/8	1	60
minim	1/61440	1/15360	1/7680	1/1920	1/480	1/60	1

Table 6 US units of volume (capacity)—dry measure only

	bu	pk	dry qt	dry pt
bushel	1	4	32	64
peck	1/4	1	8	16
dry quart	1/32	1/8	1	2
dry pint	1/64	1/16	1/2	1

Table 7 UK units of mass

	ton	cwt	(qr)	stone	lb	oz	dr
ton	1	20	80	160	2240	35 840	573 440
hundredweight	1/20	1	4	8	112	1792	28 672
quarter	1/80	1/4	1	2	28	448	7168
stone	1/160	1/8	1/2	1	14	224	3584
pound (avoir)	1/2240	1/112	1/28	1/14	1	16	256
ounce (avoir)	1/35 840	1/1792	1/448	1/224	1/16	1	16
dram (avoir)	1/573 440	1/28 672	1/7168	1/3584	1/256	1/16	1

Table 8 US units of mass

	(sh ton)	(sh cwt)	lb	oz	dr
short ton	1	20	2000	32 000	512 000
short hundredweight	1/20	1	100	1600	25 600
pound (avoir)	1/2000	1/100	1	16	256
ounce (avoir)	1/32 000	1/1600	1/16	1	16
dram (avoir)	1/512 000	1/25 600	1/256	1/16	1

Table 9 Troy units of mass

	lb t	oz tr	dwt	gr
troy pound	1	12	240	5760
troy ounce	1/12	1	20	480
pennyweight	1/240	1/20	1	24
grain	1/5760	1/480	1/24	1

Table 10 Apothecaries' units of mass

	oz apoth	dr ap	scruple	gr
apoth. ounce	1	8	24	480
drachm; dram	1/8	1	3	60
scruple	1/24	1/3	1	20
grain	1/480	1/60	1/20	1

(b) CONVERSION BETWEEN SI UNITS AND UK AND US UNITS

Table 11 Inches to millimetres

in	mm	in	mm	in	mm	in	mm	in	mm
0.001	0.0254	0.041	1.0414	0.081	2.0574	0.31	7.8740	0.71	18.0340
0.002	0.0508	0.042	1.0668	0.082	2.0828	0.32	8.1280	0.72	18.2880
0.003	0.0762	0.043	1.0922	0.083	2.1082	0.33	8.3820	0.73	18.5420
0.004	0.1016	0.044	1.1176	0.084	2.1336	0.34	8.6360	0.74	18.7960
0.005	0.1270	0.045	1.1430	0.085	2.1590	0.35	8.8900	0.75	19.0500
0.006	0.1524	0.046	1.1684	0.086	2.1844	0.36	9.1440	0.76	19.3040
0.007	0.1778	0.047	1.1938	0.087	2.2098	0.37	9.3980	0.77	19.5580
0.008	0.2032	0.048	1.2192	0.088	2.2352	0.38	9.6520	0.78	19.8120
0.009	0.2286	0.049	1.2446	0.089	2.2606	0.39	9.9060	0.79	20.0660
0.010	0.2540	0.050	1.2700	0.090	2.2860	0.40	10.1600	0.80	20.3200
0.011	0.2794	0.051	1.2954	0.091	2.3114	0.41	10.4140	0.81	20.5740
0.012	0.3048	0.052	1.3208	0.092	2.3368	0.42	10.6680	0.82	20.8280
0.013	0.3302	0.053	1.3462	0.093	2.3622	0.43	10.9220	0.83	21.0820
0.014	0.3556	0.054	1.3716	0.094	2.3876	0.44	11.1760	0.84	21.3360
0.015	0.3810	0.055	1.3970	0.095	2.4130	0.45	11.4300	0.85	21.5900
0.016	0.4064	0.056	1.4224	0.096	2.4384	0.46	11.6840	0.86	21.8440
0.017	0.4318	0.057	1.4478	0.097	2.4638	0.47	11.9380	0.87	22.0980
0.018	0.4572	0.058	1.4732	0.098	2.4892	0.48	12.1920	0.88	22.3520
0.019	0.4826	0.059	1.4986	0.099	2.5146	0.49	12.4460	0.89	22.6060
0.020	0.5080	0.060	1.5240	0.10	2.5400	0.50	12.7000	0.90	22.8600
0.021	0.5334	0.061	1.5494	0.11	2.7940	0.51	12.9540	0.91	23.1140
0.022	0.5588	0.062	1.5748	0.12	3.0480	0.52	13.2080	0.92	23.3680
0.023	0.5842	0.063	1.6002	0.13	3.3020	0.53	13.4620	0.93	23.6220
0.024	0.6096	0.064	1.6256	0.14	3.5560	0.54	13.7160	0.94	23.8760
0.025	0.6350	0.065	1.6510	0.15	3.8100	0.55	13.9700	0.95	24.1300
0.026	0.6604	0.066	1.6764	0.16	4.0640	0.56	14.2240	0.96	24.3840
0.027	0.6858	0.067	1.7018	0.17	4.3180	0.57	14.4780	0.97	24.6380
0.028	0.7112	0.068	1.7272	0.18	4.5720	0.58	14.7320	0.98	24.8920
0.029	0.7366	0.069	1.7526	0.19	4.8260	0.59	14.9860	0.99	25.1460
0.030	0.7620	0.070	1.7780	0.20	5.0800	0.60	15.2400	1.00	25,4000
0.031	0.7874	0.071	1.8034	0.21	5.3340	0.61	15.4940	1	25.4000
0.032	0.8128	0.072	1.8288	0.22	5.5880	0.62	15.7480	2	50.8000
0.033	0.8382	0.073	1.8542	0.23	5.8420	0.63	16.0020	3	76.2000
0.034	0.8636	0.074	1.8796	0.24	6.0960	0.64	16.2560	4	101.6000
0.035	0.8890	0.075	1.9050	0.25	6.3500	0.65	16.5100	5	127.0000
0.036	0.9144	0.076	1.9304	0.26	6.6040	0.66	16.7640	6	152.4000
0.037	0.9398	0.077	1.9558	0.27	6.8580	0.67	17.0180	7	177.8000
0.038	0.9652	0.078	1.9812	0.28	7.1120	0.68	17.2720	8	203.2000
0.039	0.9906	0.079	2.0066	0.29	7.3660	0.69	17.5260	9	228.6000
0.040	1.0160	0.080	2.0320	0.30	7.6200	0.70	17.7800	10	254.0000

Table 12 Millimetres to inches

mm	inches	mm	inches	mm	inches	mm	inches	mm	inches
0.01	0.000394	0.46	0.018110	0.91	0.035827	8	0.314961	17	0.66929
0.02	0.000787	0.47	0.018504	0.92	0.036220	8.2	0.322835	17.2	0.67716
0.03	0.001181	0.48	0.018898	0.93	0.036614	8.4	0.330709	17.4	0.68504
0.04	0.001575	0.49	0.019291	0.94	0.037008	8.6	0.338583	17.6	0.69291
0.05	0.001968	0.50	0.019685	0.95	0.037402	8.8	0.346457	17.8	0.70079
0.06	0.002362	0.51	0.020079	0.96	0.037795	9	0.354331	18	0.70866
0.07	0.002756	0.52	0.020472	0.97	0.038189	9.2	0.362205	18.2	0.71653
0.08	0.003150	0.53	0.020866	0.98	0.038583	9.4	0.370079	18.4	0.72441
0.09	0.003543	0.54	0.021260	0.99	0.038976	9.6	0.377953	18.6	0.73228
0.10	0.003937	0.55	0.021654	1.00	0.039370	9.8	0.385827	18.8	0.74016
0.11	0.004331	0.56	0.022047	1	0.039370	10	0.39370	19	0.74803
0.12	0.004724	0.57	0.022441	1.2	0.047244	10.2	0.40157	19.2	0.75591
0.13	0.005118	0.58	0.022835	1.4	0.055118	10.4	0.40945	19.4	0.76378
0.14	0.005512	0.59	0.023228	1.6	0.062992	10.6	0.41732	19.6	0.77165
0.15	0.005906	0.60	0.023622	1.8	0.070866	10.8	0.42520	19.8	0.77953
0.16	0.006299	0.61	0.024016	2	0.078740	11	0.43307	20	0.78740
0.17	0.006693	0.62	0.024409	2.2	0.086614	11.2	0.44094	20.0	0.79528
0.18	0.007087	0.63	0.024803	2.4	0.094488	11.4	0.44882	20.4	0.80315
0.19	0.007480	0.64	0.025197	2.6	0.102362	11.6	0.45669	20.6	0.81102
0.20	0.007874	0.65	0.025591	2.8	0.110236	11.8	0.46457	20.8	0.81890
0.21	0.008268	0.66	0.025984	3	0.118110	12	0.47244	21	0.82677
0.22	0.008661	0.67	0.026378	3.2	0.125984	12.2	0.48031	21.2	0.83465
0.23	0.009055	0.68	0.026772	3.4	0.133858	12.4	0.48819	21.4	0.84252
0.24	0.009449	0.69	0.027165	3.6	0.141732	12.6	0.49606	21.6	0.85039
0.25	0.009843	0.70	0.027559	3.8	0.149606	12.8	0.50394	21.8	0.85827
0.26	0.010236	0.71	0.027953	4	0.157480	13	0.51181	22	0.86614
0.27	0.010630	0.72	0.028346	4.2	0.165354	13.2	0.51968	22.2	0.87402
0.28	0.011024	0.73	0.028740	4.4	0.173228	13.4	0.52756	22.4	0.88189
0.29	0.011417	0.74	0.029134	4.6	0.181102	13.6	0.53543	22.6	0.88976
0.30	0.011811	0.75	0.029528	4.8	0.188976	13.8	0.54331	22.8	0.89774
0.31	0.012205	0.76	0.029921	5	0.196850	14	0.55118	23	0.90551
0.32	0.012598	0.77	0.030315	5.2	0.204724	14.2	0.55905	23.2	0.91339
0.33	0.012992	0.78	0.030709	5.4	0.212598	14.4	0.56693	23.4	0.92126
0.34	0.013386	0.79	0.031102	5.6	0.220472	14.6	0.57480	23.6	0.92913
0.35	0.013780	0.80	0.031496	5.8	0.228346	14.8	0.58268	23.8	0.93701
0.36	0.014173	0.81	0.031890	6	0.236220	15	0.59055	24	0.94488
0.37	0.014567	0.82	0.032283	6.2	0.244094	15.2	0.59842	24.2	0.95276
0.38	0.014961	0.83	0.032677	6.4	0.251969	75.4	0.60630	24.4	0.96063
0.39	0.015354	0.84	0.033071	6.6	0.259843	15.6	0.61417	24.6	0.96850
0.40	0.015748	0.85	0.033465	6.8	0.267717	15.8	0.62205	24.8	0.97638
0.41	0.016142	0.86	0.033858	7	0.275591	16	0.62992	25	0.98425
0.42	0.016535	0.87	0.034252	7.2	0.283465	16.2	0.63779	25.2	0.99213
0.43	0.016929	0.88	0.034646	7.4	0.291339	16.4	0.64567	25.4	1.00000
0.44	0.017323	0.89	0.035039	7.6	0.299213	16.6	0.65354		
0.45	0.017717	0.90	0.035433	7.8	0.307087	16.8	0.66142		

Table 13 Inches and fractions of an inch to millimetres

inches	0	1	2	3	4	5	6	7	8	9	10
0	0.00	25.40	50.80	76.20	101.60	127.00	152.40	177.80	203.20	228.60	254.00
1/64	0.40	25.80	51.20	76.60	102.00	127.40	152.80	178.20	203.60	229.00	254.40
1/32	0.79	26.19	51.59	76.99	102.39	127.79	153.19	178.59	203.99	229.39	254.79
3/64	1.19	26.59	51.99	77.39	102.79	128.19	153.59	178.99	204.39	229.79	255.19
1/16	1.59	26.99	52.39	77.79	103.19	128.59	153.99	179.39	204.79	230.19	255.59
5/64	1.98	27.38	52.78	78.18	103.58	128.98	154.38	179.78	205.18	230.58	255.98
3/32	2.38	27.78	53.18	78.58	103.98	129.38	154.78	180.18	205.58	230.98	256.38
7/64	2.78	28.18	53.58	78.98	104.38	129.78	155.18	180.58	205.98	231.38	256.78
1/8	3.18	28.58	53.98	79.38	104.78	130.18	155.58	180.98	206.38	231.78	257.18
9/64	3.57	28.97	54.37	79.77	105.17	130.57	155.97	181.37	206.77	232.17	257.57
5/32	3.97	29.37	54.77	80.17	105.57	130.97	156.37	181.77	207.17	232.57	257.97
11/64	4.37	29.77	55.17	80.57	105.97	131.37	156.77	182.17	207.57	232.97	258.37
3/16	4.76	30.16	55.56	80.96	106.36	131.76	157.16	182.56	207.96	233.36	258.76
13/64	5.16	30.56	55.96	81.36	106.76	132.16	157.56	182.96	208.36	233.76	259.16
7/32	5.56	30.96	56.36	81.76	107.16	132.56	157.96	183.36	208.76	234.16	259.56
15/64	5.95	31.35	56.75	82.15	107.55	132.95	158.35	183.75	209.15	234.55	259.95
1/4	6.35	31.75	57.15	82.55	107.95	133.35	158.75	184.15	209.55	234.95	260.35
17/64	6.75	32.15	57.55	82.95	108.35	133.75	159.15	184.55	209.95	235.35	260.75
9/32	7.14	32.54	57.94	83.34	108.74	134.14	159.54	184.94	210.34	235.74	261.14
19/64	7.54	32.94	58.34	83.74	109.14	134.54	159.94	185.34	210.74	236.14	261.54
5/16	7.94	33.34	58.74	84.14	109.54	134.94	160.34	185.74	211.14	236.54	261.94
21/64	8.33	33.73	59.13	84.53	109.93	135.33	160.73	186.13	211.53	236.93	262.33
11/32	8.73	34.13	59.53	84.93	110.33	135.73	161.13	186.53	211.93	237.33	262.73
23/64	9.13	34.53	59.93	85.33	110.73	136.13	161.53	186.93	212.33	238.73	263.13
3/8	9.53	34.93	60.33	85.73	111.13	136.53	161.93	188.33	212.73	238.13	263.53
25/64	9.92	35.32	60.72	86.12	111.52	136.92	162.32	187.72	213.12	238.52	263.92
13/32	10.32	35.72	61.12	86.52	111.92	137.32	162.72	188.12	213.52	238.92	264.32
27/64	10.72	36.12	61.52	86.92	112.32	137.72	163.12	188.52	213.92	239.32	264.72
7/16	11.11	36.51	61.91	87.31	112.71	138.11	163.51	188.91	214.31	239.71	265.11
29/64	11.51	36.91	62.31	87.71	113.11	138.51	163.91	189.31	214.71	240.11	265.51
15/32	11.91	37.31	62.71	88.11	113.51	138.91	164.31	189.71	215.11	240.51	265.91
31/64	12.30	37.70	63.10	88.50	113.90	139.30	164.70	190.10	215.50	240.90	266.30
1/2	12.70	38.10	63.50	88.90	114.30	139.70	165.10	190.50	215.90	241.30	266.70
33/64	13.10	38.50	63.90	89.30	114.70	140.10	165.50	190.90	216.30	241.70	267.10
17/32	13.49	38.89	64.29	89.69	115.09	140.49	165.89	191.29	216.69	242.09	267.49
35/64	13.89	39.29	64.69	90.09	115.49	140.89	166.29	191.69	217.09	242.49	267.89
9/16	14.29	39.69	65.09	90.49	115.89	141.29	166.69	192.09	217.49	242.89	268.29
37/64	14.68	40.08	65.48	90.88	116.28	141.68	167.08	192.48	217.88	243.28	268.69
19/32	15.08	40.48	65.88	91.28	116.68	142.08	167.48	192.88	218.28	243.68	269.08
39/64	15.48	40.88	66.28	91.68	117.08	142.48	167.88	193.28	218.68	244.08	269.48
5/8	15.88	41.28	66.68	92.08	117.48	142.88	168.28	193.68	219.08	244.48	269.88
41/64	16.27	41.67	67.07	92.47	117.87	143.27	168.67	194.07	219.47	244.87	270.28
21/32	16.67	42.07	67.47	92.87	118.27	143.67	169.07	194.47	219.87	245.27	270.67
43/64	17.07	42.47	67.87	93.27	118.67	144.07	169.47	194.87	220.27	245.67	271.07
11/16	17.46	42.86	68.26	93.66	119.06	144.46	169.86	195.26	220.66	246.06	271.46
45/64	17.86	43.26	68.66	94.06	119.46	144.86	170.26	195.66	221.06	246.46	271.86
23/32	18.26	43.66	69.06	94.46	119.86	145.26	170.66	196.06	221.46	246.86	272.26
47/64	18.65	44.05	69.45	94.85	120.25	145.65	171.05	196.45	221.85	247.25	272.65
3/4	19.05	44.45	69.85	95.25	120.65	146.05	171.45	196.85	222.25	247.65	273.05

Table 13 Continued

inches	0	1	2	3	4	5	6	7	8	9	10
49/64	19.45	44.85	70.25	95.65	121.05	146.45	171.85	197.25	222.65	248.05	273.45
25/32	19.84	45.24	70.64	96.04	121.44	146.84	172.24	197.64	223.04	248.44	273.84
51/64	20.24	45.64	71.04	96.44	121.84	147.24	172.64	198.04	223.44	248.84	274.24
13/16	20.64	46.04	71.44	96.84	122.24	147.64	173.04	198.44	223.84	249.24	274.64
53/64	21.03	46.43	71.83	97.23	122.63	148.03	173.43	198.83	224.23	249.64	275.04
27/32	21.43	46.83	72.23	97.63	123.03	148.43	173.83	199.23	224.63	250.03	275.43
55/64	21.83	47.23	72.63	98.03	123.43	148.83	174.23	199.63	225.03	250.43	275.83
7/8	22.23	47.63	73.03	98.43	123.83	149.23	174.63	200.03	225.43	250.83	276.23
57/64	22.62	48.02	73.42	98.82	124.22	149.62	175.02	200.42	225.82	251.22	276.62
29/32	23.02	48.42	73.82	99.22	124.62	150.02	175.42	200.82	226.22	251.62	277.02
59/64	23.42	48.82	74.22	99.62	125.02	150.42	175.82	201.22	226.62	252.02	277.42
15/16	23.81	49.21	74.61	100.01	125.41	150.81	176.21	201.61	227.01	252.41	277.81
61/64	24.21	49.61	75.01	101.41	125.81	151.21	176.61	202.01	227.41	252.81	278.21
31/32	24.61	50.01	75.41	100.81	126.21	151.61	177.01	202.41	227.81	253.21	278.61
63/64	25.00	50.40	75.80	101.20	126.60	152.00	177.40	202.80	228.20	253.60	279.00

Table 14 Feet to metres

ft	0	1	2	3	4	5	6	7	8	9
0	.....	0.3048	0.6096	09.144	1.2192	1.5240	1.8288	2.1336	2.4384	2.7432
10	3.0480	3.3528	3.6576	39.624	4.2672	4.5720	4.8768	5.1816	5.4864	5.7912
20	6.0960	6.4008	6.7056	70.104	7.3152	7.6200	7.9248	8.2296	8.5344	8.8392
30	9.1440	9.4488	9.7536	10.0584	10.3632	10.6680	10.9728	11.2776	11.5824	11.8872
40	12.1920	12.4968	12.8016	13.1064	13.4112	13.7160	14.0208	14.3256	14.6304	14.9352
50	15.2400	15.5448	15.8496	16.1544	16.4592	16.7640	17.0688	17.3736	17.6784	17.9832
60	18.2880	18.5928	18.8976	19.2024	19.5072	19.8120	20.1168	20.4216	20.7264	21.0312
70	21.3360	21.6408	21.9456	22.2504	22.5552	22.8600	23.1648	23.4696	23.7744	24.0792
80	24.3840	24.6888	24.9936	25.2984	25.6032	25.9080	26.2128	26.5176	26.8224	27.1272
90	27.4320	27.7368	28.0416	28.3464	28.6512	28.9560	29.2608	29.5656	29.8704	30.1752

Table 15 Metres to feet

m	0	1	2	3	4	5	6	7	8	9
0	.....	3.281	6.562	9.843	13.123	16.404	19.685	22.966	26.247	29.528
10	32.808	36.089	39.370	42.651	45.932	49.213	52.493	55.774	59.055	62.336
20	65.617	68.898	72.178	75.459	78.740	82.021	85.302	88.583	91.863	95.144
30	98.425	101.706	104.987	108.268	111.549	114.829	118.110	121.391	124.672	127.953
40	131.234	134.514	137.795	141.076	144.357	147.638	150.919	154.199	157.480	160.761
50	164.042	167.323	170.604	173.885	177.165	180.446	183.727	187.008	190.289	193.569
60	196.850	200.131	203.412	206.693	209.974	213.255	216.535	219.816	223.097	226.378
70	229.659	232.940	236.220	239.501	242.782	246.063	249.344	252.625	255.906	259.186
80	262.467	265.748	269.028	272.310	275.591	278.871	282.152	285.433	288.714	291.995
90	295.276	298.556	301.837	305.118	308.399	311.680	314.961	318.241	321.522	324.803

Table 16 Pounds to kilogrammes

lb	0	1	2	3	4	5	6	7	8	9
0	....	0.45	0.91	1.36	1.81	2.27	2.72	3.18	3.63	4.08
10	4.54	4.99	5.44	5.90	6.35	6.80	7.26	7.71	8.16	8.62
20	9.07	9.53	9.98	10.43	10.89	11.34	11.79	12.25	12.70	13.15
30	13.61	14.06	14.51	14.97	15.42	15.88	16.33	16.78	17.24	17.69
40	18.14	18.60	19.05	19.50	19,96	20.41	20.87	21.32	21.77	22.23
50	22.68	23.13	23.59	24.04	24.49	24.95	25.40	25.85	26.31	26.76
60	27.22	27.67	28.12	28.58	29.03	29.48	29.94	30.39	30.84	31.30
70	31.75	32.21	32.66	33.11	33.57	34.02	34.47	34.93	35.38	35.83
80	36.29	36.74	37.19	37.65	38.10	38.56	39.01	39.46	39.92	40.37
90	40.82	41.28	41.73	42.18	42.64	43.09	43.54	44.00	44.45	44.91
100	45.36	45.81	46.27	46.72	47.17	47.63	48.08	48.53	48.99	49.44
110	49.90	50.35	50.80	51.26	51.71	52.16	52.62	53.07	53.52	53.98
120	54.43	54.88	55.34	55.79	56.25	56.70	57.15	57.61	58.06	58.51
130	58.97	59.42	59.87	60.33	60.78	61.23	61.69	62.14	62.60	63.05
140	63.50	63.96	64.41	64.86	65.32	65.77	66.22	66.68	67.13	67.59
150	68.04	68.49	68.95	69.40	69.85	70.31	70.76	71.21	71.67	72.12
160	72.57	73.03	73.48	73.94	74.39	74.84	75.30	75.75	76.20	76.66
170	77.11	77.56	78.02	78.47	78.93	79.38	79.83	80.29	80.74	81.19
180	81.65	82.10	82.55	83.01	83.46	83.91	84.37	84.82	85.28	85.73
190	86.18	86.64	87.09	87.54	88.00	88.45	88.90	89.36	89.81	90.26
200	90.72	91.17	91.63	92.08	92.53	92.99	93.44	93.89	94.35	94.80
210	95.25	95.71	96.16	96.62	97.07	97.52	97.98	98.43	98.88	99.34
220	99.79	100.24	100.70	101.15	101.60	102.06	102.51	102.97	103.42	103.87
230	104.33	104.78	105.23	105.69	106.14	106.59	107.05	107.50	107.96	108.41
240	108.86	109.32	109.77	110.22	110.68	111.13	111.58	112.04	112.49	112.94
250	113.40	113.85	114.31	114.76	115.21	115.67	116.12	116.57	117.03	117.48
260	117.93	118.39	118.84	119.29	119.75	120.20	120.66	121.11	121.56	122.02
270	122.47	122.92	123.38	123.83	124.28	124.74	125.19	125.65	126.10	126.55
280	127.01	127.46	127.91	128.37	128.82	129.27	129.73	130.18	130.63	131.09
290	131.54	132.00	132.45	132.90	133.36	133.81	134.26	134.72	135.17	135.62
300	136.08	136.53	136.98	137.44	137.89	138.35	138.80	139.25	139.71	140.16
310	140.61	141.07	141.52	141.97	142.43	142.88	143.34	143.79	144.24	144.70
320	145.15	145.60	146.06	146.51	146.96	147.42	147.87	148.32	148.78	149.23
330	149.69	150.14	150.59	151.05	151.50	151.95	152.41	152.86	153.31	153.77
340	154.22	154.68	155.13	155.58	156.04	156.49	156.94	157.40	157.85	158.30
350	158.76	159.21	159.66	160.12	160.57	161.03	161.48	161.93	162.39	162.84
360	163.29	163.75	164.20	164.65	165.11	165.56	166.01	166.47	166.92	167.38
370	167.83	168.28	168.74	169.19	169.64	170.10	170.55	171.00	171.46	171.91
380	172.37	172.82	173.27	173.73	174.18	174.63	175.09	175.54	175.99	176.45
390	176.90	177.35	177.81	178.26	178.72	179.17	179.62	180.08	180.53	180.98
400	181.44	181.89	182.34	182.80	183.25	183.70	184.16	184.61	185.07	185.52
410	185.97	186.43	186.88	187.33	187.79	188.24	188.69	189.15	189.60	190.06
420	190.51	190.96	191.42	191.87	192.32	192.78	193.23	193.68	194.14	194.59
430	195.04	195.50	195.95	196.41	196.86	197.31	197.77	198.22	198.67	199.13
440	199.58	200.03	200.49	200.94	201.40	201.85	202.30	202.76	203.21	203.66
450	204.12	204.57	205.02	205.48	205.93	206.38	206.84	207.29	207.75	208.20
460	208.65	209.11	209.56	210.01	210.47	210.92	211.37	211.83	212.28	212.73
470	213.19	213.64	214.10	214.55	215.00	215.46	215.91	216.36	216.82	217.27
480	217.72	218.18	218.63	219.09	219.54	219.99	220.45	220.90	221.35	221.81
490	222.26	222.71	223.17	223.62	224.07	224.53	224.98	225.44	225.89	226.34

Table 17 Kilogrammes to pounds

kg	0	1	2	3	4	5	6	7	8	9
0	....	2.2	4.4	6.6	8.8	11.0	13.2	15.4	17.6	19.8
10	22.0	24.3	26.5	28.7	30.9	33.1	35.3	37.5	39.7	41.9
20	44.1	46.3	48.5	50.7	52.9	55.1	57.3	59.5	61.7	63.9
30	66.1	68.3	70.5	72.8	75.0	77.2	79.4	81.6	83.8	86.0
40	88.2	90.4	92.6	94.8	97.0	99.2	101.4	103.6	105.8	108.0
50	110.2	112.4	114.6	116.8	119.0	121.3	123.5	125.7	127.9	130.1
60	132.3	134.5	136.7	138.9	141.1	143.3	145.5	147.7	149.9	152.1
70	154.3	156.5	158.7	160.9	163.1	165.3	167.6	169.8	172.0	174.2
80	176.4	178.6	180.8	183.0	185.2	187.4	189.6	191.8	194.0	196.2
90	198.4	200.6	202.8	205.0	207.2	209.4	211.6	213.8	216.1	218.3
100	220.5	222.7	224.9	227.1	229.3	231.5	233.7	235.9	238.1	240.3
110	242.5	244.7	246.9	249.1	251.3	253.5	255.7	257.9	260.1	262.4
120	264.6	266.8	269.0	271.2	273.4	275.6	277.8	280.0	282.2	284.4
130	286.6	288.8	291.0	293.2	295.4	297.6	299.8	302.0	304.2	306.4
140	308.6	310.9	313.1	315.3	317.5	319.7	321.9	324.1	326.3	328.5
150	330.7	332.9	335.1	337.3	339.5	341.7	343.9	346.1	348.3	350.5
160	352.7	354.9	357.1	359.4	361.6	363.8	366.0	368.2	370.4	372.6
170	374.8	377.0	379.2	381.4	383.6	385.8	388.0	390.2	392.4	394.6
180	396.8	399.0	401.2	403.4	405.7	407.9	410.1	412.3	414.5	416.7
190	418.9	421.1	423.3	425.5	427.7	429.9	432.1	434.3	436.5	438.7
200	440.9	443.1	445.3	447.5	449.7	451.9	454.2	456.4	458.6	460.8
210	463.0	465.2	467.4	469.6	471.8	474.0	476.2	478.4	480.6	482.8
220	485.0	487.2	489.4	491.6	493.8	496.0	498.2	500.4	502.7	504.9
230	507.1	509.3	511.5	513.7	515.9	518.1	520.3	522.5	524.7	526.9
240	529.1	531.3	533.5	535.7	537.9	540.1	542.3	544.5	546.7	549.0
250	551.2	553.4	555.6	557.8	560.0	562.2	564.4	566.6	568.8	571.0
260	573.2	575.4	577.6	579.8	582.0	584.2	586.4	588.6	590.8	593.0
270	595.2	597.5	599.7	601.9	604.1	606.3	608.5	610.7	612.9	615.1
280	617.3	619.5	621.7	623.9	626.1	628.3	630.5	632.7	634.9	637.1
290	639.3	641.5	643.7	646.0	648.2	650.4	652.6	654.8	657.0	659.2
300	661.4	663.6	665.8	668.0	670.2	672.4	674.6	676.8	679.0	681.2
310	683.4	685.6	687.8	690.0	692.3	694.5	696.7	698.9	701.1	703.3
320	705.5	707.7	709.9	712.1	714.3	716.5	718.7	720.9	723.1	725.3
330	727.5	729.7	731.9	734.1	736.3	738.5	740.8	743.0	745.2	747.4
340	749.6	751.8	754.0	756.2	758.4	760.6	762.8	765.0	767.2	769.4
350	771.6	773.8	776.0	778.2	780.4	782.6	784.8	787.1	789.3	791.5
360	793.7	795.9	798.1	800.3	802.5	804.7	806.9	809.1	811.3	813.5
370	815.7	817.9	820.1	822.3	824.5	826.7	828.9	831.1	833.3	835.6
380	837.8	840.0	842.2	844.4	846.6	848.8	851.0	853.2	855.4	857.6
390	859.8	862.0	864.2	866.4	868.6	870.8	873.0	875.2	877.4	879.6
400	881.8	884.1	886.3	888.5	890.7	892.9	895.1	897.3	899.5	901.7
410	903.9	906.1	908.3	910.5	912.7	914.9	917.1	919.3	921.5	923.7
420	925.9	928.1	930.4	932.6	934.8	937.0	939.2	941.4	943.6	945.8
430	948.0	950.2	952.4	954.6	956.8	959.0	961.2	963.4	965.6	967.8
440	970.0	972.2	974.4	976.6	978.9	981.1	983.3	985.5	987.7	989.9
450	992.1	994.3	996.5	998.7	1000.9	1003.1	1005.3	1007.5	1009.7	1011.9
460	1014.1	1016.3	1018.5	1020.7	1022.9	1025.1	1027.4	1029.6	1031.8	1034.0
470	1036.2	1038.4	1040.6	1042.8	1045.0	1047.2	1049.4	1051.6	1053.8	1056.0
480	1058.2	1060.4	1062.6	1064.8	1067.0	1069.2	1071.4	1073.7	1075.9	1078.1
490	1080.3	1082.5	1084.7	1086.9	1089.1	1091.3	1093.5	1095.7	1097.9	1100.1

Table 18 Ounces to grammes

oz	g	oz	g	oz	g	oz	g	oz	g
1	28.35	9	255.15	17	481.94	25	708.74	40	1133.98
2	56.70	10	283.50	18	510.29	26	737.09	50	1417.48
3	85.05	11	311.84	19	538.64	27	765.44	60	1700.97
4	113.40	12	340.19	20	566.99	28	793.79	70	1984.47
5	141.75	13	368.54	21	595.34	29	822.14	80	2267.96
6	170.10	14	396.89	22	623.69	30	850.49	90	2551.46
7	198.45	15	425.24	23	652.04	31	878.84	100	2834.95
8	226.80	16	453.59	24	680.39	32	907.18		

Table 19 Grammes to ounces

g	0	1	2	3	4	5	6	7	8	9
0	....	0.04	0.07	0.11	0.14	0.18	0.21	0.25	0.28	0.32
10	0.35	0.39	0.42	0.46	0.49	0.53	0.56	0.60	0.63	0.67
20	0.71	0.74	0.78	0.81	0.85	0.88	0.92	0.95	0.99	1.02
30	1.06	1.09	1.13	1.16	1.20	1.23	1.27	1.31	1.34	1.38
40	1.41	1.45	1.48	1.52	1.55	1.59	1.62	1.66	1.69	1.73
50	1.76	1.80	1.83	1.87	1.90	1.94	1.98	2.01	2.05	2.08
60	2.12	2.15	2.19	2.22	2.26	2.29	2.33	2.36	2.40	2.43
70	2.47	2.50	2.54	2.58	2.61	2.65	2.68	2.72	2.75	2.79
80	2.82	2.86	2.89	2.93	2.96	3.00	3.03	3.07	3.10	3.14
90	3.18	3.21	3.25	3.28	3.32	3.35	3.39	3.42	3.46	3.49

Table 20 Degrees Fahrenheit to degrees Celsius

°F	0	10	20	30	40	50	60	70	80	90
0	−17.8	−12.2	− 6.7	− 1.1	4.4	10.0	15.6	21.1	26.7	32.2
100	37.8	43.3	48.9	54.4	60.0	65.6	71.1	76.7	82.2	87.8
200	93.3	98.9	104.4	110.0	115.6	121.1	126.7	132.2	137.8	143.3
300	148.9	154.4	160.0	165.6	171.1	176.7	182.2	187.8	193.3	198.9
400	204.4	210.0	215.6	221.1	226.7	232.2	237.8	243.3	248.9	254.4
500	260.0	265.6	271.1	276.7	282.2	287.8	293.3	298.9	304.4	310.0
600	315.6	321.1	326.7	332.2	337.8	343.3	348.9	354.4	360.0	365.6
700	371.1	376.7	382.2	387.8	393.3	398.9	404.4	410.0	415.6	421.1
800	426.7	432.2	437.8	443.3	448.9	454.4	460.0	465.6	471.1	476.7
900	482.2	487.8	493.3	498.9	504.4	510.0	515.6	521.1	526.7	532.2
1000	537.8	543.3	548.9	554.4	560.0	565.6	571.1	576.7	582.2	587.8
1100	593.3	598.9	604.4	610.0	615.6	621.1	626.7	632.2	637.8	643.3
1200	648.9	654.4	660.0	665.6	671.1	676.7	682.2	687.8	693.3	698.9
1300	704.4	710.0	715.6	721.1	726.7	732.2	737.8	743.3	748.9	754.4
1400	760.0	765.6	771.1	776.7	782.2	787.8	793.3	798.9	804.4	810.0
1500	815.6	821.1	826.7	832.2	837.8	843.3	848.9	854.4	860.0	865.6
1600	871.1	876.7	882.2	887.8	893.3	898.9	904.4	910.0	915.6	921.1
1700	926.7	932.2	937.8	943.3	948.9	954.4	960.0	965.6	971.1	976.7
1800	982.2	987.8	993.3	998.9	1004.4	1010.0	1015.6	1021.1	1026.7	1032.2
1900	1037.8	1043.3	1048.9	1054.4	1060.0	1065.6	1071.1	1076.7	1082.2	1087.8
2000	1093.3	1098.9	1104.4	1110.0	1115.6	1121.1	1126.7	1132.2	1137.8	1143.3
2100	1148.9	1154.4	1160.0	1165.6	1171.1	1176.7	1182.2	1187.8	1193.3	1198.9
2200	1204.4	1210.0	1215.6	1221.1	1226.7	1232.2	1237.8	1243.3	1248.9	1254.4
2300	1260.0	1265.6	1271.1	1276.7	1282.2	1287.8	1293.3	1298.9	1304.4	1310.0
2400	1315.6	1321.1	1326.7	1332.2	1337.8	1343.3	1348.9	1354.4	1360.0	1365.6

°F	1	2	3	4	5	6	7	8	9	10
°C	0.6	1.1	1.7	2.2	2.8	3.3	3.9	4.4	5.0	5.6

Table 21 Degrees Celsius to degrees Fahrenheit

°C	0	10	20	30	40	50	60	70	80	90
0	32	50	68	86	104	122	140	158	176	194
100	212	230	248	266	284	302	320	338	356	374
200	392	410	428	446	464	482	500	518	536	554
300	572	590	608	626	644	662	680	698	716	734
400	752	770	788	806	824	842	860	878	896	914
500	932	950	968	986	1004	1022	1040	1058	1076	1094
600	1112	1130	1148	1166	1184	1202	1220	1238	1256	1274
700	1292	1310	1328	1346	1364	1382	1400	1418	1436	1454
800	1472	1490	1508	1526	1544	1562	1580	1598	1616	1634
900	1652	1670	1688	1706	1724	1742	1760	1778	1796	1814
1000	1832	1850	1868	1886	1904	1922	1940	1958	1976	1994
1100	2012	2030	2048	2066	2084	2102	2120	2138	2156	2174
1200	2192	2210	2228	2246	2264	2282	2300	2318	2336	2354
1300	2372	2390	2408	2426	2444	2462	2480	2498	2516	2534
1400	2552	2570	2588	2606	2624	2642	2660	2678	2696	2714
1500	2732	2750	2768	2786	2804	2822	2840	2858	2876	2894
1600	2912	2930	2948	2966	2984	3002	3020	3038	3056	3074
1700	3092	3110	3128	3146	3164	3182	3200	3218	3236	3254
1800	3272	3290	3308	3326	3344	3362	3380	3398	3416	3434
1900	3452	3470	3488	3506	3524	3542	3560	3578	3596	3614
2000	3632	3650	3668	3686	3704	3722	3740	3758	3776	3794
2100	3812	3830	3848	3866	3884	3902	3920	3938	3956	3974
2200	3992	4010	4028	4046	4064	4082	4100	4118	4136	4154
2300	4172	4190	4208	4226	4244	4262	4280	4298	4316	4334
2400	4352	4370	4388	4406	4424	4442	4460	4478	4496	4514

°C	1	2	3	4	5	6	7	8	9	10
°F	1.8	3.6	5.4	7.2	9.0	10.8	12.6	14.4	16.2	18.0

(c) LATEST DEFINITIONS OF SI BASE-UNITS

THE METRE is the length equal to 1 650 763.73 wavelengths in vacuum of the radiation corresponding to the transition between the levels $2p_{10}$ and $5d_5$ of the krypton-86 atom (XI CGPM, 1960).

THE KILOGRAMME is equal to the mass of the international prototype of the kilogramme (III CGPM, 1901).

THE SECOND is the duration of 9 192 631 770 periods of the radiation corresponding to the transition between two hyperfine levels of the ground state of the caesium-133 atom (XIII CGPM, 1967).

THE AMPERE is that constant current which, if maintained in two straight parallel conductors of infinite length, of negligible circular cross-section, and placed 1 metre apart in a vacuum, would produce between these conductors a force equal to 2×10^{-7} newton per metre of length (IX CGPM, 1948).

THE KELVIN is the fraction 1/273.16 of the thermodynamic temperature of the triple point of water (XIII CGPM, 1967).

THE CANDELA is the luminous intensity, in the perpendicular direction, of a surface of 1/600 000 square metre of a black body at the temperature of freezing platinum under a pressure of 101 325 newtons per square metre (XIII CGPM, 1967).

Bibliography

ANDERTON, P. and BIGG, P. H. *Changing to the metric system*, London, HMSO, 3 edn., 1969.

ASTM Metric practice guide, National Bureau of Standards, Handbook 102, 1967.

BEIGBEDER ATIENZA, F. *Conversiones metrológicas entre los sistemas Nortamericano, Inglés, Métrico Decimal, Cegesimal y Giorgi*, Madrid, 1952.

CHISHOLM, L. J. *Units of weight and measure*, National Bureau of Standards, Miscellaneous Publication 286, 1967.

EBERT, H. *Physikalisches Taschenbuch*, Braunschweig, Vieweg, 4 edn., 1967.

FRANKE, H. *Lexikon der Physik*, 3 vol., Stuttgart, Franckh'sche Verlagshandlung, 3 edn., 1969.

HVISTENDAHL, H. S. *Engineering units and physical quantities*, London, Macmillan, 1964.

JERRARD, H. G. and MCNEILL, D. B. *A dictionary of scientific units*, London, Chapman Hall, 2 edn., 1964.

KAYE, G. W. C. and LABY, T. H. *Tables of physical and chemical constants*, edited by committee, London, Longmans Green, 13 edn., 1966.

KONEČNÝ, L. *ABC technických jednotiek*, Technická práca, 1963.

MESTON, LORD. *Shaw's Guide to the Weights and Measures Act* 1963, London, Shaw, 1964.

SACKLOWSKI, A. *Einheiten der Physik und Technik*, Stuttgart, Deutsche Verlags-Anstalt, 3 edn., 1966.

Symbols, units and nomenclature in physics, Document U.I.P. 11 (S.U.N. 65–3), IUPAP, 1965.

Symbols, signs and abbreviations, London, Royal Society, 1969.

ŠINDELÁŘ, V. and SMRŽ, L. *Nová měrová soustava v ČSSR*, Normalisace, 1962–4.

YOUNG, L. *System of units in electricity and magnetism*, Edinburgh, Oliver and Boyd, 1969.

British Standards

BS 350 Conversion factors and tables
- Part 1, Basis of tables. Conversion factors
- Part 2, Detailed conversion factors

BS 1637 Memorandum on the M.K.S. system of electrical and magnetic units

BS 1991 Letter symbols, signs and abbreviations
- Part 1, General
- Part 2, Chemical engineering, nuclear science and applied chemistry
- Part 3, Fluid mechanics
- Part 4, Structures, materials and soil mechanics
- Part 5, Applied thermodynamics
- Part 6, Electrical science and engineering

BS 2990 Rationalized and unrationalized formulae in electrical engineering

BS 3763 International System (SI) units

French Standards

FD X 02–002	Unités de mesure—Définitions
FD X 02–006	Unités de mesure—Choix d'unités usuelles
FD X 02–007	Unités de mesure—Aide-mémoire pour l'application du Système International (SI)
FD X 02–010	Sous-multiples décimaux du degré
NF X 02–050	Principales unités de mesure américaines et britanniques

German Standards

DIN 1301	Einheiten (+ Entwurf)
DIN 1304	Allgemeine Formelzeichen
DIN 1305	Masse, Gewicht, Gewichtskraft, Fallbeschleunigung
DIN 1306	Dichte
DIN 1311	Schwingungslehre (Bl. 1, Entw.)
DIN 1314	Druck
DIN 1315	Winkeleinheiten, Winkelteilungen

DIN 1320 Aknstik. Grundbegriffe
DIN 1324 Elektrisches Feld (Entw.)
DIN 1325 Magnetisches Feld
DIN 1332 Akustik. Formelzeichen (+ Bbl. 1)
DIN 1339 Einheiten magnetischer Grössen
DIN 1341 Wärmeübertragung
DIN 1342 Viskosität bei Newtonschen Flüssigkeiten (+ Entw.)
DIN 1343 Normzustand, Normvolumen
DIN 1344 Formelzeichen der elektrischen Nachrichtentechnik
DIN 1345 Technische Thermodynamik (+ Bl. 10 Entw.)
DIN 1349 Strahlungsdurchgang durch Medien (Bl. 1 Entw.)
DIN 1350 Zeichen für Festigkeitsberechnungen
DIN 1355 Zeit
DIN 1357 Einheiten elektrischer Grössen
DIN 5031 Strahlungsphysik im optischen Bereich und Lichttechnik
DIN 5491 Stoffübertragung (Entw.)
DIN 5492 Formelzeichen der Strömungsmechanik
DIN 5493 Logarithmierte Verhältnisgrössen
DIN 5494 Grössensysteme und Einheitensysteme
DIN 5496 Temperaturstrahlung
DIN 5497 Mechanik; Starre Körper
DIN 5498 Chemische Thermodynamik
DIN 24 315 Ölhydraulik und Pneumatik; Einheiten-Vergleich
DIN 28 400 Vakuumtechnik; Fachausdrücke und Begriffe (Bl. 1 Entw.)
DIN 45 630 Grundlagen der Schallmessung (Bl. 1)

International Electrotechnical Vocabulary (IEC50)

Group 07, Electronics, 1956
Group 08, Electro-acoustics, 1960
Group 26, Nuclear power plants for electric energy generation, 1968
Group 65, Radiology and radiological physics, 1964
Group 66, Detection and measurement of ionizing radiation by electric means, 1968

ISO Recommendation R 31
Part I, Basic Quantities and Units of the SI (and Quantities and Units of Space and Time), 1965
Part II, Quantities and units of Periodic and Related Phenomena, 1958
Part III, Quantities and Units of Mechanics, 1960
Part IV, Quantities and Units of Heat, 1960
Part V, Quantities and Units of Electricity and Magnetism, 1965
Part VII, Quantities and Units of Acoustics, 1965

ISO Recommendation R 131, Expression of the physical and subjective magnitudes of sound or noise, 1959

ISO Recommendation R 357, Expression of the power and intensity levels of sound or noise, 1963

ISO Recommendation R 1000, Rules for the use of units of the International System of Units and a selection of the decimal multiples and sub-multiples of the SI units, 1969

Draft ISO Recommendation No. 838, Quantities and Units of atomic and nuclear physics, 1965

Draft ISO Recommendation No. 839, Quantities and Units of nuclear reactions and ionizing radiations, 1965

Draft ISO Recommendation No. 1777, Quantities and Units of physical chemistry and molecular physics, 1969

Draft ISO Recommendation No 1778, Quantities and Units of light and related electromagnetic radiations, 1969

French Index

This index sets out the French names of all quantities and constants. Names of units have been included only where they differ from the English.

German Index

This index sets out the German names of all the quantities and constants. Names of units have been included only where they differ from the English.

Important factors derived from pi

$\pi \approx 3.141\ 592\ 653\ 589\ 793$

2π	6.283 185	$1/\pi^2$	$1.013\ 212 \times 10^{-1}$
3π	9.424 778	$1/\pi^3$	$3.225\ 153 \times 10^{-2}$
4π	$1.256\ 637 \times 10$	$1/\pi^4$	$1.026\ 598 \times 10^{-2}$
$\pi/2$	1.570 796	$1/\pi^5$	$3.267\ 764 \times 10^{-3}$
$\pi/3$	1.047 198	$\sqrt{\pi}$	1.772 454
$4\pi/3$	4.188 790	$\sqrt{(2\pi)}$	2.506 628
$\pi/4$	$7.853\ 982 \times 10^{-1}$	$\sqrt{(2/\pi)}$	$7.978\ 846 \times 10^{-1}$
$\pi/6$	$5.235\ 988 \times 10^{-1}$	$\sqrt{(3/\pi)}$	$9.772\ 050 \times 10^{-1}$
$\pi\sqrt{2}$	4.442 883	$\sqrt{(4/\pi)}$	1.128 379
$\pi/\sqrt{2}$	2.221 442	$\sqrt{(\pi/2)}$	1.253 314
$1/\pi$	$3.183\ 099 \times 10^{-1}$	$\pi\sqrt{\pi}$	5.568 328
$1/2\pi$	$1.591\ 549 \times 10^{-1}$	$\sqrt[3]{\pi}$	1.464 592
$1/4\pi$	$7.957\ 747 \times 10^{-2}$	$\sqrt[3]{(2\pi)}$	1.845 270
$2/\pi$	$6.336\ 198 \times 10^{-1}$	$\sqrt[3]{(2/\pi)}$	$8.602\ 540 \times 10^{-1}$
$3/\pi$	$9.549\ 297 \times 10^{-1}$	$\sqrt[3]{(3/\pi)}$	$9.847\ 450 \times 10^{-1}$
$4/\pi$	1.273 240	$\sqrt[3]{(6/\pi)}$	1.240 701
π^2	9.869 604	$\sqrt[3]{(3/4\pi)}$	6.203 505
π^3	$3.100\ 628 \times 10$	$\sqrt[3]{(\pi/2)}$	1.162 447
π^4	$9.740\ 909 \times 10$	$\sqrt[3]{(\pi/4)}$	$9.226\ 351 \times 10^{-1}$
π^5	$3.060\ 197 \times 10^2$	$\sqrt[3]{(\pi/6)}$	$8.059\ 960 \times 10^{-1}$
$4\pi^2$	$3.947\ 842 \times 10$	$\sqrt[3]{(\pi^2)}$	2.145 029
$\pi^2/4$	2.467 401	$\pi \times \sqrt[3]{\pi}$	4.601 151
		$1/\sqrt[3]{\pi}$	$6.827\ 841 \times 10^{-1}$
		$1/\sqrt{\pi}$	$5.641\ 896 \times 10^{-1}$

Table of decimal equivalents of selected fractions

1/2	$5.000\,00 \times 10^{-1}$	1/128	$7.812\,50 \times 10^{-3}$
1/3	$3.333\,33 \times 10^{-1}$	1/144	$6.944\,44 \times 10^{-3}$
1/4	$2.500\,00 \times 10^{-1}$	1/176	$5.681\,82 \times 10^{-3}$
1/5	$2.000\,00 \times 10^{-1}$	1/192	$5.208\,33 \times 10^{-3}$
1/6	$1.666\,67 \times 10^{-1}$	1/198	$5.050\,51 \times 10^{-3}$
1/7	$1.428\,57 \times 10^{-1}$	1/224	$4.464\,29 \times 10^{-3}$
1/8	$1.250\,00 \times 10^{-1}$	1/254	$3.937\,01 \times 10^{-3}$
1/9	$1.111\,11 \times 10^{-1}$	1/256	$3.906\,25 \times 10^{-3}$
1/10	$1.000\,00 \times 10^{-1}$	1/288	$3.472\,22 \times 10^{-3}$
1/12	$8.333\,33 \times 10^{-2}$	1/448	$2.232\,14 \times 10^{-3}$
1/14	$7.142\,86 \times 10^{-2}$	1/512	$1.953\,12 \times 10^{-3}$
1/16	$6.250\,00 \times 10^{-2}$	1/528	$1.893\,94 \times 10^{-3}$
1/18	$5.555\,56 \times 10^{-2}$	1/576	$1.736\,11 \times 10^{-3}$
1/20	$5.000\,00 \times 10^{-2}$	1/768	$1.302\,08 \times 10^{-3}$
1/22	$4.545\,45 \times 10^{-2}$	1/792	$1.262\,63 \times 10^{-3}$
1/24	$4.166\,67 \times 10^{-2}$	1/1024	$9.765\,63 \times 10^{-4}$
1/26	$3.846\,15 \times 10^{-2}$	1/1089	$9.182\,74 \times 10^{-4}$
1/27	$3.703\,70 \times 10^{-2}$	1/1296	$7.716\,05 \times 10^{-4}$
1/28	$3.571\,43 \times 10^{-2}$	1/1536	$6.510\,42 \times 10^{-4}$
1/32	$3.125\,00 \times 10^{-2}$	1/1728	$5.787\,04 \times 10^{-4}$
1/36	$2.777\,78 \times 10^{-2}$	1/1792	$5.580\,36 \times 10^{-4}$
1/48	$2.083\,33 \times 10^{-2}$	1/3048	$3.280\,84 \times 10^{-4}$
1/54	$1.851\,85 \times 10^{-2}$	1/3584	$2.790\,96 \times 10^{-4}$
1/64	$1.562\,50 \times 10^{-2}$	1/4356	$2.295\,68 \times 10^{-4}$
1/66	$1.515\,15 \times 10^{-2}$	1/6144	$1.627\,60 \times 10^{-4}$
1/96	$1.041\,67 \times 10^{-2}$	1/6336	$1.578\,28 \times 10^{-4}$
1/112	$8.928\,57 \times 10^{-3}$	1/7168	$1.395\,09 \times 10^{-4}$
1/121	$8.264\,46 \times 10^{-3}$		

Most Common Symbols of Units and Their Prefixes

A	ampere
a	are; ab-
at	atmosphere (technical)
atm	atmosphere (physical)
b	barn; (bar); (bel)
Btu	British thermal unit
C	coulomb
°C	degree Celsius
c	centi-
cal	calorie (I.T.)
cd	candela
cm	centimetre
d	day; deci-
dB	decibel
dm	decimetre
dyn	dyne
erg	erg
F	farad
°F	degree Fahrenheit
ft	foot
g	gramme
gal	gallon
gr	grain
H	henry
h	hour; hecto-
ha	hectare
Hz	hertz
in	inch
J	joule
K	kelvin
°K	degree Kelvin
k	kilo-
kcal	kilocalorie I.T.
kg	kilogramme
kgf	kilogramme-force
km	kilometre
kmol	kilomol
kp	kilopond
l	litre
lb	pound
lbf	pound-force
lm	lumen
lx	lux
M	mega-
m	metre; milli-
min	minute; minim
mm	millimetre
mol	mole
N	newton
oz	ounce (avoirdupois)
pdl	poundal
pt	pint
rad	radian
S	siemens
s	second; stat-
sr	steradian
T	tesla
t	tonne
V	volt
W	watt
Wb	weber
yd	yard
μ	micron; micro-
Ω	ohm
....°	degree (of angle)
....′	minute (of angle)
....″	second (of angle)